INSURGENCY ONLINE:
WEB ACTIVISM AND GLOBAL CONFLICT

In *Insurgency Online*, Michael Dartnell focuses on a new form of conflict made possible by today's communication systems. The Internet, Dartnell argues, is effecting extensive changes to the way politics is carried out by allowing a range of non-state actors onto the global political stage. He demonstrates that Web activism raises issues about the organization of societies and the distribution of power, and contends that the development of online activism has far-reaching social and political implications with parallels to the influence of the invention of the printing press, the telegraph, and the radio.

Dartnell concentrates on Web activists who use the Internet as a media tool, distinguishing this use from information terrorism, which threatens or harasses through 'hacking' or electronic sabotage. Using the examples of the Revolutionary Association of the Women of Afghanistan (RAWA), which opposed the Taliban, the Peruvian Movimiento Revolucionario Tupac Amaru (MRTA) and its campaign against the Fujimori government, and the Irish Republican Socialist Movement (IRSM), Dartnell evaluates the political implications and general character of Web activism. *Insurgency Online* shows that online activism is a ripe, new method for non-governmental movements and individuals to raise awareness and develop support around the world.

(Digital Futures)

MICHAEL Y. DARTNELL is an associate professor in the Department of History and Politics at the University of New Brunswick, Saint John.

Insurgency Online

Web Activism and Global Conflict

Michael Y. Dartnell

UNIVERSITY OF TORONTO PRESS
Toronto Buffalo London

© University of Toronto Press Incorporated 2006
Toronto Buffalo London
Printed in Canada

ISBN-13: 978-0-8020-8747-8 (cloth)
ISBN-10: 0-8020-8747-7 (cloth)

ISBN-13: 978-0-8020-8553-5 (paper)
ISBN-10: 0-8020-8553-9 (paper)

Printed on acid-free paper

Library and Archives Canada Cataloguing in Publication

Dartnell, Michael
    Insurgency online : web activism and global conflict / Michael Y.
Dartnell.

    Includes bibliographical references and index.
    ISBN-13: 978-0-8020-8747-8 (cloth)
    ISBN-13: 978-0-8020-8553-5 (pbk.)
    ISBN-10: 0-8020-8747-7 (cloth)
    ISBN-10: 0-8020-8553-9 (pbk.)

    1. Internet – Political aspects.   2. Internet – Social aspects.
    3. Subversive activities.   4. Political activists.   I. Title.

    HM851.D36 2006   303.48'33   C2005-906035-2

University of Toronto Press acknowledges the financial assistance to its
publishing program of the Canada Council for the Arts and the Ontario
Arts Council.

University of Toronto Press acknowledges the financial support for its
publishing activities of the Government of Canada through the Book
Publishing Industry Development Program (BPIDP).

To Martin Chochinov,
for support, encouragement, advice, insight, endless dinners, dancing
and ideas, movies, glamour, style, wit, passion ... a life.

'It's a casserole, silly!'

Contents

Foreword

Insurgency Online is critical media theory for the twenty-first century.

Refusing a passive approach to understanding media, *Insurgency Online* does something more challenging and, in the end, far more rewarding. Intently studying genuinely revolutionary changes in the nature of mass media initiated by the massive, still-accelerating growth of electronic culture and the Internet, *Insurgency Online* explores the impact of the Internet upon political resistance. Notice here that we are not in the old world of politics existing somehow outside the daily life of the media nor in the now-superseded world of mass media with its implicit assumption that politics is somehow at a critical remove from media.

In *Insurgency Online*, we are immediately taken through the looking-glass of the Internet to the very real world of electronic politics. When we step through the pages of the text we discover a world of clashing politics which exists *only in and by means of* the electronic media of communication. In the industrial age, we experienced the politics of industrialism with its elite-driven parties and mass political organizations of voters. In the age of the Internet, politics migrates from its top-down, hierarchical, one-way, passive industrial form, becoming something fluid, networked, participatory, immediate, resistant, democratic. Industrial-strength politics might make kings of ruling elites, opinion makers of editorial writers, with access to mass media tightly controlled by the power of money, but Internet politics is strikingly different. It provides a global voice to radically excluded groups. In a way that Marshall McLuhan predicted long ago, it signals a future politics in which there are no centres, only margins. Necessarily democratic and resistant and fluid in its method of electronic diffusion, Internet politics is perfect for the age of *singularities* – the era in which what counts is not the size of the consumer audience but the strength of the political signal. Before we ever arrive at the actual content of electronic politics, we have already passed over to a new style of political communication which is insurgent in its very form and which, like a rebellion from below, from above, from just about everywhere, puts out a strong signal of authentic political struggle that will not be denied, although its presence will definitely go unreported by mass media, which, not surprisingly, have everything to lose by the wild popularity of politics online.

Unreported, that is, until the publication of *Insurgency Online*. Here, dispensing with the normal rules of political interpretation which hold that analysis should remain distinct from and always faithfully representational of its volatile subject matter, the Canadian political scientist Michael Dartnell has the great insight to do McLuhan one better – to actually analyse politics from *within* the Web, to immerse critical thought in the flood of media images, to study how the structure of electronic media actually shapes perception, to listen intently to the clash of ideological perspectives, to make a pioneering meditation on power and domination in the (hyper)real world of electronic politics.

> The flood of images is what I call 'insurgency online.' Over the past ten years, new information technology (IT) in general and the Web in particular have transformed, accelerated, and extended the representational and image-based dimensions of conflict. While the twentieth century episodically witnessed the rise of the moral and emotional appeals of propaganda on radio and television, through photographs, and in movie theatres, a significant minority of the global population now receives a constant stream of images and text at work and in the privacy of home. This discussion focuses on the identarian and political aspects of this change by examining Web activism, a form of global conflict made possible by the World Wide Web.

Directly challenging international relations research on the politics of the Internet, *Insurgency Online* articulates a suspicion awakening in 'post-realist' political researchers everywhere these days, namely that media analysis focused on public institutions such as the 'state' and private institutions such as 'corporate governance' are no longer adequate in themselves as living signs of actual political experience. When politics is streamed by images, when mass emotions are crisis-driven by the spikes and dips of fast-flowing information, when politics begins to speak with an Internet accent, we know that we are moving at the speed of light. We are in the world that Einstein prophesied would be a 'spacetime' fabric held together by shifting, highly relative viewpoints – a world that is both subject and object of *Insurgency Online*.

Consequently, more than a brilliant study of Web activism, *Insurgency Online* represents a major challenge to received interpretations of post–Cold War politics. Written at a time when the dust has finally cleared from the tearing down of the Berlin Wall (and, with it, what Russian President Putin has recently described as the 'global catastrophe' of the collapse of Soviet-style communism), and mindful of the quick movement of contemporary political culture to globalized militarism in the wake of 9/11, *Insurgency Online* represents a highly creative, pioneering attempt to think anew the world of twenty-first-century politics. Here, realist approaches to international relations which valorize the role of the state give way to an *actual* global political experience where insurgent forces of nationalism and religion have multiplied political identities, fragmented old Cold War binaries, and destabilized the meaning of ideology, conflict, and communication. As Dartnell notes, the shift in communication symbolized by the proliferation of the Web, by the global buzz of the Internet, occurs simultaneously with the migration of politics from 'hard power' to a new world of 'soft power,' where issues related to economics, culture, technology, and identity are paramount.

> Web media are part of a shift in politics that could be as far-reaching, profound, and unpredictable as the rise of print technology, mass literacy, and nationalism in the late eighteenth century.

Perfect. Because, following in the critical tradition of the great Canadian communication theorist Harold Innis, Michael Dartnell sets out to say something very serious, very challenging about world politics, viewed through the creative lens of Web activism. Faithful to his epistemological perspective on the migratory, fluid, destabilized nature of international politics post-9/11, *Insurgency Online* plunges directly into the real world of politics today. What are politely introduced as 'non-state' actors quickly are revealed to be the nodal points of contemporary political conflict – Kurdish rebels, Ukrainian civic nationalists, Zapatista revolutionaries, Chechnyan warriors. Finally, for once, we have a method of understanding international politics which links together authentically revolutionary developments in mobile electronic technologies with equally revolutionary political movements, quick to take advantage of new technologies of electronic communication for their own communicative purposes. In *Insurgency Online*, media theory grows directly out of the contested ground of the hard politics of actual media politics.

A creative method of media analysis as well as an equally critical method of political analysis, *Insurgency Online* focuses specifically on three important examples of Web activism: in Europe, the Irish Republican Socialist Movement (IRSM); in Asia, the Revolutionary Association of the Women of Afghanistan (RAWA); and in Latin America, the Movimiento Revolucionario Tupac Amaru (MRTA). At a time when political analysis of the changing landscape of international politics is actively discouraged by the information apparatus of the 'War against Terrorism,' Dartnell's Web-focused research does exactly the opposite. Paying close attention to the political use of Web activism by Irish Republicans, feminists in Afghanistan, and Peruvian revolutionaries, his analysis breaks the vow of silence surrounding real political insurgencies in the twenty-first century. To read these chapters with their explicit media analysis of the ups and downs of Web activism, to follow Dartnell's research as it traces the rise and fall of political movements through diverse information strategies is actually to be taking, perhaps quite unexpectedly, a political journey deep into the heart of contemporary political conflicts. Like all meaningful journeys, this is not one where the reader returns unchanged, untouched by this brush with the stubborn, recalcitrant, unswerving rebellion of the Irish Republican Socialist Movement, the struggle of Afghan women against the Taliban, or the militancy of the Tupac Amaru. While *Insurgency Online* focuses on nationalist, feminist, and leftist movements, its truly creative method of new media analysis could as easily be applied to other political movements operating under the sign of IT: faith-based, conservative, (electronic) tribalist. As the title of the final chapter states so succinctly, Web activism is a 'messenger that shapes perceptions.'

Arthur Kroker
Canada Research Chair in Technology, Culture, and Theory
and Director of the Pacific Centre for Technology and Culture
at the University of Victoria

Acknowledgments

Listen up, people. This book is a result of work that I started in 1997 on the Web. The research and writing that I carried out in Montreal, Toronto, Windsor, and Saint John have benefited from many individuals and several institutions. It is a great pleasure to be able to name and thank those who enabled me to complete the text.

David Scott, now Chief Engineer at NetZentry in Cupertino, California, sparked the idea for *Insurgency Online*. David's brilliant and wide-ranging imagination, his humour, and his straightforward way of talking about tech opened a new world for me. Thanks David.

I would like to thank the U.S. Institute for Peace (USIP), the NATO Fellowship program, and the York University Centre for International and Security Studies (YCISS). Each of these organizations provided critical assistance that have made the book possible. A generous USIP grant in 1999–2000 gave me the time to research, write, and think about the Web as well as the funds to hire research assistants. At that time, I was a Research Associate at YCISS, a place where I have consistently benefited from strong support. I would particularly like to thank David Dewitt, David Mutimer, Joan Broussard, Sarah Whitaker, Heather Chestnut, and the other YCISS faculty and staff, whose combined work made for an exciting and stimulating environment. David Dewitt and David Mutimer (the 'Davids') both provided a secure intellectual environment while I did my work. While at YCISS, I was also very fortunate to have the help of a group of research assistants who immeasurably aided my work. Thanks to Sofie Tzoutzi for her excellent research on the RAWA website, Geoff Kennedy for his superb work on Northern Ireland and the IRSM, as well as Julie Senna, Marnie Lucas-Zerbe, and Ananda Chakravarty.

A number of friends and colleagues have discussed and challenged the ideas and research presented here, among them: Shannon Bell, Gad Horowitz, Bob Whitney, Debbie Payne, Leslie Jeffrey, Don Desserud, Robin Sutherland, Joanna Everitt, Lindsay Bell, Henry Habib, Bruce Tucker, Martha Lee, and Ian Spears. I would also especially like to thank Marilouise and Arthur Kroker, whose friendship and example have inspired me since we met at the University of Winnipeg longer ago than any of us probably care to remember.

Other comments and suggestions have been provided as the text was written. In this regard I would like to thank Marcus Schulz, Saskia Sassen, Paul Wilkinson, Louise Richardson, John Hogan, Ron Deibert, Michael Cox, Chris Hables Gray, James Der Derian, and others who, at various times in different contexts, offered their insights on portions of the research that went into the book.

Finally, thanks to my wonderful editor at the University of Toronto Press, Siobhan McMenemy. Siobhan is a real pro and a wise person, whose insights and foresight really led the way to the text that you are holding. Thanks again Siobhan.

INSURGENCY ONLINE

1 Introduction: Insurgency Online and Conflict in the Global-scape

As the second Gulf War raged in March and April 2003, a struggle over perceptions of the conflict erupted in newspapers, cable television networks, and on the World Wide Web. Television viewers in the West repeatedly saw images of murals of Saddam Hussein being bulldozed, shot at, or otherwise obliterated. Photographs of Saddam were shown being smashed, shot at, and removed from prominent positions in public places in Iraq. A particularly hot spot in the battle of images was the Qatar-based Arabic-language television news network Al Jazeera, whose broadcast of U.S. prisoners captured by Saddam's forces outraged the Pentagon and supporters of the Anglo-American campaign. Al Jazeera senior editor Faisal Bodi noted that 'one measure of the importance of those American PoW pictures and the images of the dead British soldiers is surely the sustained "shock-and-awe" hacking campaign directed at aljazeera.net since the start of the war.'[1] Hacking – or, more precisely, distributed denial of service (D DOS) – attacks flooded the Al Jazeera website with up to 300 megabits (Mbps) of data per second. The hacking of the Al Jazeera website paralleled the battle over images, representations, and identities that each set of combatants brought to the war. Like the battle over images, hacking took advantage of the communicative capacities of the Internet platform. Hacking is a technical challenge that provokes deep-seated anxieties over the impact of technology, which stem from a 'Frankenstein' model in which science and technology are feared as having unintended, dangerous, and grotesque consequences. The struggle over perceptions on the Web is both more subtle and flagrant than hacking. It occurs through a flood of images, values, interests, and ideas that shock, titillate, challenge, and reinforce. It also embodies, in a world of microbes, tissue manufacture, suicide bombings, kidnapping, and genocide, another transformation in the conceptions, meanings, and conduct of politics and conflict.

The flood of images is what I call 'insurgency online.' Over the past ten years, new information technology (IT) in general and the Web in particular have transformed, accelerated, and extended the representational and image-based dimensions of conflict. While the twentieth century episodically witnessed the rise of the moral and emotional appeals of propaganda on radio and television, through photographs, and in movie theatres, a significant minority of the global population now receives a constant stream of images and text at work and in the privacy of home. This discussion focuses on the identarian and political aspects of this change by examining Web activism, a form of global conflict made possible by the World Wide Web. Non-state actors around the globe transgress state-produced and -regulated text, images, and representation by means of Web activism. Through this, they subtly alter the boundaries that are the foundations of conventional territorial politics. The transformation has import for politics, security, and conflict because boundary transgression means that the values, interests, and needs that underlie personal and group identities as well as states are being re-articulated and re-negotiated. Rather than propaganda, Web activism is a practice through which notions of self, community, the relationship of self to community, the links between self, community, and technology, and participatory practices explicitly enter conflict.

Web activism has been made possible by the development of a global mediascape. At the same time, Web activism substantiates the global mediascape, though only partially. The global mediascape refers

> both to the distribution of the electronic capabilities to produce and disseminate information (newspapers, magazines, television stations, film production studios, etc.), which are now available to a growing number of private and public interests throughout the world; and to the images of the world created by these media. These images of the world involve many complicated inflections, depending on their mode (documentary or entertainment), their hardware (electronic or pre-electronic), their audiences (local, national or transnational) and the interests of those who own and control them. What is most important about these mediascapes is that they provide (especially in their television, film and cassette forms) large and complex repetoires of images, narratives and 'ethnoscapes' to viewers throughout the world, in which the world of commodities and the

world of 'news' and politics are profoundly mixed. What this means is that many audiences throughout the world experience the media themselves as a complicated and interconnected repertoire of print, celluloid, electronic screens and billboards.[2]

As Arjun Appadurai notes, the mediascape can be a place where the ties between the real and the fictional are blurred. It thus becomes a forum in which individuals and groups build 'imagined worlds' that are the foundations of power and politics. In a world in which states are relatively de-centred for a variety of reasons, mediascapes serve as important bases from which to fabricate ways of looking and doing.

In this discussion, the Irish Republican Socialist Movement (IRSM), the Revolutionary Association of the Women of Afghanistan (RAWA), and the Movimiento Revolucionario Tupac Amaru (MRTA or Tupac Amaru) illustrate the mediascape, Web activist practice, and some of their possible impacts. In spite of their distance from each other and their very different histories, all three organizations are strongly marked by the failure of the state, that mainstay of realist views of global politics. By transgressing the territorial boundaries of political society and identity, Web activism brings IT and media from the background of conflict to centre stage. The transgressive contents of Web activism raise issues about how human societies are organized, how power is distributed, and how global communication has been transformed in the past few decades. These issues have ramifications that resemble the impact of the printing press, the invention of the photograph and telegraph, and the rise of radio and television. In spite of this, international relations (IR) research on the politics of the Internet has largely ignored images and messages because it focuses on the security of large states and corporate institutions or nostalgically longs for individual privacy. With disarming speed, the Web has transformed how we communicate and express our identities and values. Or has it? Instead of print images focused on territorial units called states, contemporary global politics is a transnational, '24/7' exchange of text, photos, audio and video clips, blogs, and chat rooms that constantly transmits and re-transmits the emotional and moral content of politics. But have emotions and moralities been changed? The answer in this text will frustrate those looking for clear models of politics, conflict, and security since the Web alters ways of communicating (practices) without directly changing contents. To paraphrase Marshall McLuhan, the Web is a messenger that shapes perceptions, not events.

Insurgency Online explores this new arena of global conflict by concentrating on how non-state actors use the Web. Specifically, the focus is on how RAWA, MRTA, and IRSM use the Web to provide information in campaigns against particular governments. Each organization understands the potential of the Web environment and, in different ways, uses it to place issues before global civil society. In this respect, Web activism is a conceptual tool to highlight a variant of so-called 'information terrorism,' evaluate the political implications of IT, and analyse the Web's features as a medium. The aim is to sharpen the focus between non-state actors that use Web activism in conflict and others who threaten or harass institutions or individuals through 'hacking' or electronic sabotage. In this way, the relevance of the mediascape for IR will be shown. Hacking is unlike Web activism but often amalgamated with it because both practices occur in an electronic setting.[3] However, hacking, which is also called 'website defacement,' is here understood as tampering with or distorting information. The impact of hacking is often overstated since information management now employs very large networks as a way to circumvent tactics such as D DOS attacks. In contrast, Web activism centres on producing, providing, and spreading information outside of government control or regulation. To illustrate the new political world that emerged following IT dissemination, *Insurgency Online* examines feminist, nationalist, and leftist non-state actors with distinct individual applications of Web activism.

A Culture and Technology Approach to Analysis of the Web in International Relations

Although the Web transforms global communication in several ways, its impact must be understood in relation to the available knowledge, skills, and infrastructure that shape how, or if, it is accessed by users. The Web is an environment that provides interactivity, synchronicity, flexibility, affordability, and availability to a limited yet rapidly changing population. It functions independently of its physical location. Provocatively, the medium is used by previously unknown, marginal, or illegal non-state actors in specific jurisdictions to articulate views, values, and goals, forming 'complex non-territorial-based links that defy the organization of political authority in the modern world.'[4] Non-state actors that were once largely limited to specific spaces are now actively *transnational* due to Web media. At the same time, issues that were once contained within territorial space have now become global issues.

In spite of dazzling connectivity and capacities, the impact of the Web is very relative since the skills, knowledge, and infrastructure necessary for successful use of the technology are not evenly distributed around the globe or between countries. This is readily seen in statistics on various world regions and countries. In 2002, the International Telecommunication Union (ITU) said there were 99 Internet users for every 10,000 people in Africa, 2421 in the Americas (North and South), 557 in Asia, 2079 in Europe, and 3330 in Oceania. The African average was about one-tenth of the world average of 972 users per 10,000 persons. In 2002, there were 1.23 personal computers (PCs) per 100 inhabitants in Africa, 27.5 in the Americas, 3.95 in Asia, 20.01 in Europe, and 38.94 in Oceania. The access gap is even more striking when individual countries are compared. In 2002, the United States had 62.5 PCs for every 100 persons and 5375 Internet users per 10,000 inhabitants. Meanwhile, there were .05 PCs per 100 persons in Niger and 1.14 Internet users for every 10,000 inhabitants of the Democratic Republic of Congo.[5] These discrepancies suggest that the impact of the Web is not always direct if it is significant at all. For the majority of the planet, recent technological innovations have had no impact on knowledge, infrastructure, or skills.

There are several streams of interpretation of IT use by non-state actors. Some observers argue that IT 'de-territorializes' politics by diminishing the need to base political interaction in physical locations. The above data on Internet access and PC use suggest that IT-driven de-territorialization may only be true, if at all, for a minority of the world's inhabitants. This view is further substantiated by ITU data on cell phone and telephone use, which resemble those on the Internet and PCs insofar as Africans and Asians have much less access than their peers in Europe, Oceania, or the Americas.[6] Other analysts view IT as inherently dangerous. Studies of information terrorism emphasize how IT directly threatens national and international security. This line of interpretation grew from the literature on political terrorism in which authors such as Brigitte Lebens Nacos contend that media access must be limited to strengthen security and order.[7] These views are based in hyperbole and threaten the goal of a relatively open global civil society.[8] *Insurgency Online* differs from either approach because it interprets the Web within wider post–Cold War transformations rather than as part of a bipolar confrontation such as that between the USSR and the United States from 1945 to 1989. While polarized politics between right and left characterized the Cold War era, in the post–Cold War period polarization no longer determines every conflict. In the 1990s, the struggle between capitalism and communism was supplanted by the multiplication of political identities seen in today's nationalist and religious movements. Contending and often contradictory political claims that increasingly come into contact in unexpected ways are a significant factor in this global setting.

This book explores how the Web fits into the post–Cold War world by integrating theories of media and communication, comparative politics, international relations, and security studies. The discussion blends an analysis of non-state actors[9] with an interpretation of the political impact of the Web through an interdisciplinary approach to IR called 'technology and culture.' The approach assesses Web activism as a practice that seeks to shape perceptions by using technology to place ideas, values, and beliefs before global civil society. It views values and their expression as motivating forces in human societies. The approach goes beyond the struggle to survive that preoccupies materialist interpretations of IR in order to examine what differentiates human beings from others species: culture and technology. The culture and technology analysis demonstrates that the practice of Web activism has varying impacts. IRSM Web activism gave coherence to a geographically dispersed network of individuals and organizations. RAWA Web activism documented allegations of gender abuse in order to influence global public opinion. MRTA Web activism involved relaying information about the group and its methods to mainstream media. In spite of individual specificities, all three cases were marked by an effort to reshape perceptions that mixes the value-laden cultural and practical-technological components of contemporary communication devices.

Insurgency Online also addresses the theoretical implications of the 'physical force' tradition of politics in an era in which conflict is as much about perceptions as it was once about securing defined territory. The term 'physical force tradition of politics' comes from analyses of radical Irish nationalism, especially with reference to the various forms of Irish republicanism in the twentieth century. The tradition is based on a belief that only violent direct action will achieve a meaningful transformation of society. I suggest that perceptions are as important as violence in political and social change because putting values into practice preoccupies human societies. IT augments the role of perceptions by increasing the availability of information and enhancing our means of communication. An underlying assumption below is that communication is now relatively more important to global conflict than in the Cold War era because contemporary IT circumvents the once-direct state regulation of telegraphs, radio, and television. Communication now occurs globally through satellites, the Internet, cell phones, cable news networks, and radio, all of which make content control much more difficult for governments. The difficulty in controlling information and its impact was obvious in the aftermath of 9/11, when acts of organized global violence by non-state actors became media events that derailed and redefined politics and policy in the United States and elsewhere.[10]

By distinguishing technology used for information provision by non-state actors from information terrorism, *Insurgency Online* sets out new analytical categories and clarifies issues for future research. The text defines some of the phenomenological contents of Web activism (media relay, global witnessing, and networking) that move analysis away from generalization toward an assessment of particular impacts and implications. It also examines how Web activism is a fulcrum through which the impact of culture in combination with technology can be evaluated. In so doing, the discussion casts the political implications of globalized IT into a long-term, structural perspective while providing guidelines for more immediate purposes. The chapters on select movements in Europe (the IRSM), Asia (RAWA), and Latin America (MRTA) locate them and outline their character in relation to their social, political, and global environments. The aim is to show that information has a context of meaning that cannot be ignored. There are a variety of non-state Web activists whose goals are neither entirely threatening nor destructive.

Insurgency Online interprets electronic and contextual fieldwork and builds a framework that links technology and culture to contemporary conflict.[11] A number of questions underlie this interpretation: What is the character and significance of the online 'communities' manifest in Web activism? Does Web communication affect the relationship between various governments and insurgent oppositions? Do Web activists threaten the state and its ability for autonomous action (sovereignty)? Are the relationships, connections, and communities spawned by Web activism in any way analogous to the varieties of nationalism shaped by rising literacy and print technology after the eighteenth century? The chapters on IRSM, RAWA, and the MRTA substantively address Web activism as a way through which non-state actors take aim at perceptions and transgress the boundaries of identities and political societies. The conclusion in all three cases is that Web activism does not threaten the state or its ability for autonomous action, but that (and this is key) it transforms the contexts in which states operate when considered alongside factors such as multinational corporations, non-governmental organizations, migration, environmental issues, transnational criminal networks, epidemics such as AIDS and SARS, and other factors. By examining how information by non-state actors shapes perceptions, this text defines an aspect of global conflict that demonstrates why physical control of territory is only one component of legitimate governance.

Policy-makers now confront a proliferation of groups that use the Web for different ends. In an era in which two anonymous individuals at a distance can enter a chat room using an alias and discuss (for example) planting a bomb in a public place with relative ease, the security implications of IT are clear. The difficulties of monitoring such activities are daunting, most likely beyond the logistical scope of intelligence organizations as well as social science. However, if social sciences were limited to what is empirically verifiable, they would indeed be narrow and irrelevant practices. Given that IT enhances human abilities for communication, there *is* a great deal of difference between meeting in an alleyway and meeting in a chat room since human communication is considerably transformed by overcoming the limits of physical contiguity. As such, distinguishing information-provision from public and/or security threats substantially enhances our capacity to characterize and qualify threats. It helps distinguish longer-term or structural concerns from issues that demand immediate response. In terms of civil rights and freedom of expression issues, differentiating Web activism from information terrorism clarifies issues around the possibility and desirability of regulating online expression. While information terrorism interferes with communication, Web activism conveys information that governments, elites, or portions of global civil society might find inconvenient, uncomfortable, or offensive. The issue is whether annoying or inconvenient information should be limited. In the event, the regulation of Web activism occurs in a global and transnational context.

Insurgency Online examines nationalist IRSM, feminist RAWA, and leftist MRTA as select samples of the broad range of African, Asian, European, Latin American, and North American non-state actors present on the Web. English-language materials were used in order to examine how non-state actors address global civil society in a leading international language. The discussion analyses the contents of these materials and the manner in which goals are communicated. The significance of the materials is examined from an interdisciplinary perspective that combines historical analysis of message contexts, ideological analysis of the character of each group (for example, whether nationalist, religious, or another viewpoint), technical analysis of message transmission (for example, text, photographs, video, or multimedia), and discourse analysis (e.g., what goals are expressed and what symbols are used).

The dissemination and use of Web technology raise critical interpretative issues for understanding the trajectories and processes of contemporary global conflict. Students of social movements, for example, need to understand how Web activists use technology to organize across time and distance. Analysts of activism will want to assess how Web activism influences direct action movements and tactics and examine its success and failures. Researchers who focus on non-state actors will want to assess the impact of Web activism on global politics. Policy-makers will want to examine how IT changes the conditions for state action, response, and intervention. Social science research on the uses of the Web in conflict is especially needed due to the spread of regular, irregular, and often unregulated communication by non-state actors.[12]

activism @ info.web

The impact of information in the contemporary world is difficult to precisely measure even though its role in international economics and global society as well as in shaping the conditions for power is clear.[13] The U.S. Department of Commerce is very precise as to the economic value of information:

> The dollar value of e-commerce and the e-commerce share of economic activity vary markedly among key economic sectors ... Manufacturing leads all industry sectors with e-commerce shipments that account for 18.4 percent ($777 billion) of the total value of manufacturing shipments ... Merchant Wholesalers rank second with e-commerce sales that represent 7.7 percent ($213 billion) of their total sales ... Retail Trade, the focus of much e-commerce attention, has e-commerce sales in 2000 that account for 0.9 percent ($29 billion) of total retail sales.[14]

No regulations and few standard measures exist to qualify the assertion. There is no systematic measure of the economic impact of information by international agencies that would facilitate assessment of a global political economy of information. Although analysts such as Dan Schiller have completed theoretical work that conceptualizes 'information' as an economic factor, it has not yet been transformed into a framework for measurement. In the absence of clearly quantifiable measures, assessing the qualitative impact of information on society, culture, and politics is even more difficult.

Analyses of the relationship between IT and conflict must be based in frameworks that facilitate understanding. However, there is little analytical consensus over the definition of 'information,' which is a term that can be used in several ways. The volume of available information has greatly increased and differentiated into many sectors as a result of innovations in IT. Information can be used to measure the organization and coherence of various sensory data about the world. Information is also a commodity that is bought and sold when 'data mining,' for example, packages it for particular business or state clients. Finally, information also represents the world, encapsulating experience into specific sequences. The wide variety of information means that comprehensively measuring its multiple impacts in scientific innovation, marketing, accounting, news, ads, propaganda, education, publishing, trading, computer hardware and software, electronic sales, and other areas of political, economic, cultural, and social activity is not easy.

An example of the difficulty in understanding the significance of information is a 1999 statement on electronic commerce by the U.S. Bureau of the Census.[15] The text admits that the information economy is largely undefined and unrecognized in official economic statistics. It acknowledges that establishing terms to clearly and consistently describe a growing and dynamic network economy is critical for developing useful statistics about it, but adds that the terms Internet, electronic commerce, electronic business, and cybertrade are interchangeable, with no common understanding of their scope or relationships. Not surprisingly, conceptual vagueness and sheer breadth also hamper the discussion of IT in global politics and conflict.[16] It is difficult to talk about the relationship between IT and conflict in the absence of a common language about what is being discussed. Nor is it clear how to characterize the significance of the communities linked to IT, which are less geographical than other political communities. This discussion of Web activism aims to help clarify issues by referring to information as a measure that structures identity and political boundaries.

In IR, understanding the global impact of IT is hobbled by the influence of realist theory, which centres on the notion that states and the principle of sovereignty shape the international system and provide a structure to contain the chaos of human diversity. This school of IR argues that non-state actors such as corporations, international organizations such as the UN, and transnational movements (e.g., Greenpeace or the movement to ban anti-personnel land mines) operate in a context that is primarily shaped by intergovernmental relations. For realists, the impact of IT is measured in relation to states, their institutions and regulations, and the patterns of inter-state ties. The framework does not easily incorporate an analysis of a global-scape in which IT has a relatively autonomous influence on how issues are presented and understood. The examples of IRSM, RAWA, and MRTA Web activism highlight sets of practices that have relative autonomy from the state due to their ability to secure specific effects in the international system and because they have emerged in postcolonial settings in which state formation is incomplete, weak, or deeply flawed.

Realist theories of IR do not provide a suitable framework for examining the IT that emerged over the past thirty years because the communication environment now extends beyond states. For most of the twentieth century, information flowed through national postal systems and state-regulated telecommunication networks. During the twentieth century, state-based structures designed for snail mail and telegraphs gradually incorporated new media such as radio and TV. Electronic communication still occurred as one-way transmission from sender to receiver. Only in the last quarter of the twentieth century did information begin to flow in two or more directions through a decentralized and multi-layered global system in which satellite television, cell phones, fax machines, email, and the Web greatly eased cross-border and inter-hemispheric telecommunication by individuals and groups. As a result, a significant minority of the world's population routinely sends and receives messages and immediately accesses information from other parts of the globe.[17] The situation cannot be accounted for within a conceptual framework (realism) that was developed to explain power in relatively static, competing territorial entities.

Several approaches that explain the influence of IT in global politics go beyond the realist preoccupation with states. *International regime theorists* such as Mark Alleyne examine the legal-institutional framework for international communication. In Alleyne's case, the focus is the domination by media in the North over those in the South.[18] Theories based in *international political economy* treat information as a commodity. Dan Schiller, for example, discusses the need to understand information as something produced instead of a resource inherent in the nature of the world.[19] *Neo-liberal institutionalists* such as Robert Keohane and Joseph Nye, Jr, concentrate on a concept of complex interdependence that treats information as one influence among others, including states, international organizations, economic forces, and social movements.[20] *Peace and conflict theories* focus on the transformation of the global-scape and political identities due to development of IT and the collapse of the bipolar Cold War order.[21] *Global communications theory* examines IT as the latest form of international communication alongside maps, telegraph, postal service, and other media.[22] Finally, *media theory* argues that IT structures representations of the world and focuses on the role of images in shaping power in contemporary societies.[23]

All of these theories view innovations in IT and the media they carry as part of a general shift in political and global life.[24] The shift is generally cast as a move away from the high politics of foreign affairs and defence, which Nye and Keohane call the 'hard power' of force, coercion, and the military. They argue that the world is moving toward a form of politics that enhances the visibility of local issues and the 'soft power' of economics, culture, and technology. Given the brutal eruption of violence around 9/11 and continued war in the Middle East, Central Asia, Africa, Europe, and Latin America, this assessment is at first glance puzzling. The overall narrative is increasingly complex since IT gives non-state actors the means to violently strike at global civil society and governments as well as to influence and engage them. In this setting, Web activists spread information, attempt to alter perceptions, and embody the complexity of global power. Alongside military and cybersecurity IT applications, Web activism illustrates how global conflict has changed since the 1980s. Web activism would have once been dismissed as 'propaganda' if it were episodic and local-physical but now occurs within a complex and changing global mediascape. Web activists directly engage perceptions by more fully articulating a worldview, highlighting how security has become entwined with how we view the world.

IT-driven change is a security concern for states. As early as 1995, a report to the U.S. Assistant Secretary of Defense noted that Internet use by domestic and international political movements played a key role in Desert Storm, the Tianamen Square massacre, the attempted coup in Russia, the conflict in the former Yugoslavia, and in the challenge to authoritarian controls in Iran, China, and other oppressive states. The Internet is playing an increasingly significant role in international security.[25]

The assessment acknowledges the security implications of dramatically increased volumes of information and diminished time-distance constraints. Some effects were demonstrated by anti-globalization protests in Seattle, Prague, and Quebec City. Others are evident in the Web activities of groups such as Al Qaeda, organizations such as Human Rights Watch (HRW) and Doctors Without Borders, and the anti-occupation movements in Palestine and Iraq.[26]

Web media are part of a shift in politics that could be as far-reaching, profound, and unpredictable as the rise of print technology, mass literacy, and nationalism in the late eighteenth century. After 1789, territorial units called states dominated politics, replacing the web of local, religious, and dynastic power that characterized early modern Europe. In principle, the state had perpetual and absolute power or 'sovereignty' within specific boundaries.[27] The concept of sovereignty was originally developed to give order and coherence to political life. Although sovereignty is *not* central to this discussion, the self-directing capacities of territorially defined political units are fundamental referents for contemporary understandings of politics and IR. A range of contemporary practices that include IT now challenge sovereignty. In this discussion, 'sovereignty' follows Monroe Price's definition of 'the power of a state (or other accumulation of power) to make and enforce laws and to seek to have a monopoly of the use of force.'[28] From 1917 to 1989, the world was divided into camps based on communism, capitalism, and, at times, fascism. Division of the world into spatial entities based on ideological camps overlaid the anxiety of territorial power, which were seen in Cold War maps in which communism was depicted as seeping into the rest of world from the former Soviet Union and China. States used sovereignty to regulate and organize information via postal services, telecommunication regulations, educational standards, public records, political symbols, and, sometimes, ownership of media outlets such as newspapers and TV networks. The system of spatial-territorial power shaped a form of political identity in which 'where I am' decisively shaped 'who I am.'

Global political identities and ideologies have fragmented along lines that include gender, ethno-nationalism, religion, left- and right-wing agendas, sexuality, language, and region. The Web provides the wide range of non-state actors representing these identities with an opportunity to present themselves in a global-scape. These identities and associations are shaped in a world in which physical location is suddenly less important and many voices are heard. Ambivalence, doubt, influence, and truth-value all orient analysis of the IT-IR relationship toward a form of complexity that has been variously described. Saskia Sassen argues 'the ascendance of an international human rights regime and of a large variety of nonstate actors in the international arena signals the expansion of an international civil society.'[29] James Rosenau concurs that 'an endless series of distant proximities in which the forces pressing for greater globalization and those inducing greater localization interactively play themselves out.'[30] Appadurai speaks of 'the complexity of the current global economy [that] has to do with certain fundamental disjunctures between economy, culture, and politics that we have only begun to theorize.'[31] In this world, Web media facilitate global power and are 'essential to the projection of influence and the mobilization of public opinion.'[32] The IT–media–global power relationship needs more study to highlight complexity and its theoretical implications.

In territorial politics, images and text in pamphlets, maps, flags, anthems, language, radio, and TV represented state power. They gave coherence to what Benedict Anderson calls the 'imagined communities' of nations.[33] In post–Cold War politics, the Web and other globalizing IT facilitate articulation of the emotions of identity and belonging as well as a sense of morality over greater distance, at lower cost, with more computing power, and with vastly increased amounts of information. Kurdish nationalists adapted IT to the needs of their far-flung diaspora through the satellite television station MedTV as well as mailing lists, news, information, history, culture, music, language, images, and video accessed on the Web.[34] In an early example of non-Western Web activism, the Web carried a global Kurdish response to the arrest of Kurdish Workers' Party (PKK) leader Abdullah Ocalan in 1998. From an orchestrated online base, protests swept west from Kazakhstan to Vancouver in widespread and rapid response that would have once been beyond the Kurds' logistical, organizational, and financial grasp.

The Kurdish example pointed to how the Web alters the conditions for understanding and interpreting power. The limits to this Web activism are shaped by cost and culture. Cost influences include the expense of access, equipment, and the education needed to carry out Web activism. Culture is influential because Web use is linked to a pre-disposition to use electronic devices, perceptions of its utility, efficacy, and relevance,[35] and the eventual impact of the information that is transmitted. Dorothy Denning notes that in this context, 'security requires an awareness of potential attacks and constant vigilance. Even so, it is an elusive goal, never fully attainable and always competing with the desire for flexibility, access, and performance. Ultimately, security is about risk management.'[36]

The Web levels the playing field as much as it empowers players. While the threat of tampering with information or interfering with crit-ical infrastructures – variously called website defacement, cyberterror-ism, or information warfare – is real, the Web and other IT also vastly increase security and intelligence capacities. This was seen in the quick detection of Real IRA cell phone use after the 1998 Omagh bombing in Northern Ireland. The attackers were arrested because police traced a borrowed cell phone that was used in the attack.

In relative terms, Web activism diminishes the centrality of force while it increases the influence of complex social and political relation-ships in conflict.[37] Displacing coercion is an important transformation because global power has traditionally been linked to a state's tangible military resources. The greater influence of soft-power politics based in convincing, appealing to, attracting, legitimizing, and encouraging makes possible a form of conflict that is increasingly based in percep-tions. An example of this is the Ukrainian Orange Revolution in November-December 2004, in which parts of the old left had trouble rec-onciling 'revolution' with liberal democratic demands, many of them established in the West. The Ukrainian perception, in contrast, was that these demands were a legitimate form of progress for their society. In this complex international setting, Web activism is a springboard for shaping *perceptions* in a global-scape in which states physically control areas but cannot regulate information to citizens in once-routine ways. In the Second World War, Allied and Axis powers essentially shut out counter-arguments. In a world of Web activism, states interfere, but even determined efforts by Chinese and Saudi governments do not completely eliminate the entry of what they consider dangerous views via the Internet. The entry of information can be circuitous:

The Internet (including chat rooms and e-mail) has been widely used by

Middle Easterners living or working abroad, including for discussions of political issues in their home countries. As these expatriates return home, or as local users participate in the same online forums, the use of the Internet for political discourse in a more liberal environment could have an effect on the way the medium is used at home.[38]

Perceptions are integral to conflict as never before since the legitimacy of government actions and politics is more open to scrutiny and criticism (in the West, with the exception of post-9/11 America). An important concern in this landscape is the quality, quantity, and uses made of information as well as access to sophisticated understandings of IT. Historically, states shaped and controlled political information. The Nazi-backed film *Triumph of the Will*, for example, structured information to spread a message through sound, images, and emotions. Seventy years later, the Web provides decentralized two-way communication based in a culture of technology in which groups adapt to and adopt new media for various reasons (such as age, gender, education, values, or location).

Web activism embodies the increased complexity of the post–Cold War mediascape in several ways. First, political and technological elites use the Web to gain global attention. Among the alternative elites, the Zapatistas were some of the first to use the Web to establish a global profile and solidify their stance vis-à-vis the Mexican government. Second, the Web transnationalizes local and national conflict insofar as certain sets of issues arouse increasing concern in other parts of the world.[39] Even a conflict as distant from traditional Western concerns as Chechnya has a greater profile. Third, transnational movements such as the anti-personnel land mine movement, human rights organizations, white supremacists, environmentalists, and The Vatican use the Web to enhance their visibility and flexibility. Finally, societies experience anxiety due to the introduction of new political concerns and/or because of increased information about existing issues. In the 1930s, the Western left was largely in the dark about Stalin's starvation of Ukrainians. In 2004, the movement that coalesced around Viktor Yushchenko was followed in detail. Web activism is a type of electronic direct action that might aim to: confound beliefs about a society, culture, and government; strike fear and perhaps disorient people; aim to appeal, convince, and attract; and work to legitimize ideas, values, and organizations. Web activism's power lies in its strongly symbolic, image-driven, and representational form along with its claim to provide immediate and authentic information from specific groups.

Insurgency Online and Conflict in a Post-Realist Global-scape

The nationalist, feminist, and leftist Web activists discussed here highlight the diversity of non-state actors. They are presented to help develop analytic criteria and methodology to evaluate Web activist messages. This text evaluates their success, their use of Web technology, and the social context of their information (political issues, type of organization – democratic or non-democratic – and nature of the conflict in question). The focus is conflict for several reasons. In the early twenty-first century, conflict is a central metaphor that links disparate groups and contexts. In this case, it links three contexts and groups. Governments' ability to provide security and non-state actors' capacity to challenge is at the heart of politics and conflict. A central justification for power by particular elites is their ability to structure individual, group, and social life on predictable, productive, and peaceful foundations. Since the three groups here do not threaten states per se or provide parallel structures of governance, characterizing their behaviour is a key issue in evaluating the Web's impact. Each group is involved in a conflict or active in relation to an environment in which conflict has been ignited. Web activism only involves conflict in a traditional sense insofar as it concerns transgressing the boundaries for identity and political society.

The contemporary global mediascape is a setting in which anyone with a computer can communicate globally, institutions have less control, and states cannot completely regulate information flows. It is now possible to globally disseminate information with a low-cost computer, modem, and Internet connection. Despite this and the implied power to destabilize boundaries, IT does not overwhelm states so much as it alters relations between physical locations. Media use points to a big change in the political development of information: '[The] information revolution implies the rise of cyberwar, in which neither mass nor mobility will decide outcomes; instead the side that knows more, that can disperse the fog of war yet enshroud an adversary in it, will enjoy decisive advantages.'[40]

In 2004–5, U.S. action in Iraq, which involves waging a highly publicized war while inadvertently enshrouding the 'enemy' in sympathetic global media coverage, seems destined to fail. Some of this paradox can be traced to the political impact of IT, since increased decentralization and globalized information transmission make 'it possible for diverse, dispersed actors to communicate, consult, coordinate, and operate together across greater distances, and on the basis of more and better information than ever before.'[41] In terms of the Iraq conflict, the U.S. finds itself in a militarily superior position, but more friendless. In this global-scape, the perceptions that underlie politics change since the information, ideas, and networks to sustain identities and societies are themselves transformed.

The impact of the Web can be seen in the proliferation of messages, the emergence of new voices, and the development of new elites.[42] The Web increases non-state actors' ability to organize and produce messages to disseminate locally or globally.[43] The Web is not undesirable or dangerous to the stability of national and global politics, but raises new concerns and ways of conceiving politics such as the increased prominence of Islamic ideas in domestic political debates in Canada, the UK, the United States, France, and elsewhere. This suggests that in a mediascape power flows to those who edit and validate information and sort out what is correct and significant. The Web could eventually develop into a tool for political lobbying as TV did after the 1950s (seen in the Kennedy-Nixon debate). Some Web activism might aim to confound beliefs about the nature of society, culture, and government to the end of striking fear and perhaps disorienting people while other forms will also aim to appeal, convince, and attract as well as legitimize ideas, values, and movements.[44]

The three non-state actors in this discussion were chosen because of their links to well-known conflicts in Northern Ireland (civil conflict since the 1960s), Afghanistan (foreign invasion and civil war), and Peru (guerrilla insurgency). They reflect the Web's political relevance in very different societies in physically separate locations: a Western society with a long-standing history of violent conflict; a Latin American society marked by incomplete state formation and weak national identity that was challenged by indigenous and left-wing alternatives as well as extreme social and political violence; and a Central Asian society characterized by foreign invasion, civil war, and failure to develop peaceful methods of conflict resolution. The examples show how the Web has truly global significance that cuts across conventional political assumptions. In Afghanistan, as noted below, the population is not even online. In Northern Ireland, the constituency that is addressed online resides in the United States, the UK, Australia, and elsewhere where diaspora communities exist, as well as on the island of Ireland. Foreign supporters in remote locations conducted MRTA Web activism. Their efforts did not prevent the slaughter of the guerrillas in Peru. All three examples show how Web activism features varied actors (feminist, nationalist, leftist) with distinct relations to their context and different aims. In this book, a focused comparison method is used for examining different types of groups in distinct locales by concentrating on their common feature: Web activism.

Each website here is analysed on the basis of a 'snapshot,' that is, an examination of its contents at a particular point between 1997 and 2000. A snapshot approach was used due to the difficulties in examining a variable medium. Websites are designed to change over time in response to needs. This quality is underlined by the fact that, by mid-2003, the MRTA website was not available in English, although Spanish, Serbo-Croatian, Italian, and Japanese-language sites were still available.[45] The MRTA website located at BURN! no longer exists. The USA Patriot Act of 2001 made it an offence to provide 'material support or resources' to foreign terrorist organizations and effectively curtailed the online presence of some radical groups at BURN![46] This shows how Web activism faces severe government regulation in some contexts. The disappearance of the English-language MRTA website underlines the importance of timely social science examinations of specific Web activists. The Web continues to provide outlets through chat rooms and applications such as multimedia files in which illegal content is more difficult to detect. A consequence of increased government surveillance, then, is use of variable forms of website communication, which was seen in the use of multimedia files by Islamic extremists in 2003.[47] As such, the fate of the English-language MRTA website illustrates the *relative* power of the Web vis-à-vis states.

The relative power of the Web is based in its distinct capacities. One quality of Web technology is its ability to hypermediate, that is, incorporate existing media such as photos, video, radio, and even TV.[48] This is an example of how the Web embodies Chris Hables Gray's view of information as a central weapon, myth, metaphor, force multiplier, edge, trope, factor, and asset because it shows how technology can be adapted to various narrative formats.[49] Both the form and content of the Web put the public face-to-face with a complex global-scape of politics, culture, and technology. Since the real advantages of decentralized information and new hierarchies in this setting go to those with access, skills, and networks,[50] Web activists are a privileged group.

They benefit from high speed, low costs, asynchronicity, multiple-user communication, automation, and intelligent applications. The privileged position of Internet use is seen in India, which benefited greatly from the information boom in the 1990s. The BBC reports that

> India's dazzling performance in IT has been hailed as a great hope for the country's future development. The industry is one of India's fastest-growing sectors, its software analysts have become a prestigious export in themselves and India is a centre for overseas data processing from accounts to customer calls. However, just a small proportion of educated people have access to IT – but the vast majority of Indians, about 70% of the population, still live in villages and the challenge is to make sure they don't get left behind.[51]

The issue is what Web activists do with their privileged global situation. The medium provides an opportunity to improve organization, efficiency, recruitment, and morale for a host of sexual, religious, ethnic, ethical, environmental, economic, and cultural causes. Since lower cost and fewer government controls are only available to groups that access computer networks, another form of inequality has emerged.[52] Reflecting their distinct characters as groups and capacities of the medium, the MRTA, RAWA, and IRSM each exploit the Web differently. The MRTA successfully relayed a campaign against the Fujimori government into mainstream media in the West and elsewhere, and used the website to build a global support network. RAWA used the Web in the same way, but its main achievement was to document gender abuse in Afghanistan. IRSM Web activism improved organizational cohesion and dialogue as well as contact between physically distant components of the movement. For all three non-state actors, the medium facilitates an identity that is 'no longer dependent on a territorial community (*Gemeinschaft*) or on formal organizations (*Gesellschaft*) but on networks (*Verbindungnetzschaft*).'[53]

Of course, the Internet also structures behaviours and expectations. Authority exists in the forms of moderated mailing lists, chatrooms that shape relations, Internet service providers (ISPs) that set contractual conditions for subscribers, technical know-how, and efforts at national regulation. Since Web activists usually aim to shape behaviour and elicit responses rather than annihilate society, they often conform to online norms so that they can reach a public. Elaborate and expensive website management, operation, and financing are beyond the capacities of many non-state actors. The IRSM, RAWA, and MRTA sites incorporate sophisticated interactivity through email, hyperlinks, and images for support groups around the world.

Through these diverse practices, Web activism demonstrates how the once-exclusive control of governments and well-financed interests over media and communication has been transformed. Small organizations and individuals with Web pages and computer access now 'reinforce the cosmopolitanism of the new professional and managerial classes living symbolically in a global frame of reference, unlike most of the population in any country.'[54] The relative availability of low-cost ISPs means that small and even unpopular non-state actors participate in a new electronic narrative whereas previously 'they were often frozen out of the traditional media, and their views often were not taken seriously.'[55] Search engines that index the Web and use keywords to locate sites or information about movements and organizations enhance the reach of non-state actors. Websites usually provide links to other similar organizations. The MRTA and RAWA websites, for example, archive domestic and foreign news coverage of their activities. At the same time, Web activism reflects the social conditions of small non-state actors since, by comparison with the electronic presence of wealthy organizations such as the U.S. Republican National Committee, the MRTA, RAWA, and IRSM websites are relatively unsophisticated.

One practice that helps smaller groups overcome this disadvantage is email. Email is one of the most visible features of the Internet and is central to Web communication. Email is a practice based in two-way communication that can be adapted for multiple users. The Internet's vast email systems, online services, company computers, personal data assistants (PDAs), and network servers provide small groups with rapid, easy, and cheap opportunities to reach a wired public. Email can be a flexible tool that is accessed through desktop, hand-held, and wireless devices. Email transmits simple texts, word-processing files, images, and sound, video, or database materials to individuals or multiple recipients on mailing lists. Activist email can directly contact media with press releases while urgent information is easily handled, filed, and/or retrieved. Communication also occurs semi-privately through automated mailing lists or 'list servers' that re-transmit messages to recipients. List servers are not entirely private, but do overcome some of the time and distance barriers to internal debate and routine information. Email can also be risky for non-state actors. In an age of endless spam, activist email might annoy a public or lead to government regulation if it is unwelcome, persistent, or seen as extreme. Unsolicited email[56] could lead recipients to set their systems to refuse mail from specific senders.[57] Third parties might also receive, change, or re-transmit messages in an altered form[58] with the aim of seriously discrediting a group or cause. Many people also might not read email from a list because they already receive 'too much' information. Like email virus transmission, this is an unintended blowback effect of eased communication and huge volumes of information on the Internet.

In spite of these limits, Web use allows non-state actors to move beyond passive (i.e., one-way) radio-television transmission and adopt two-way communication. The Web does not access a traditional mass public, but rather a network of potential participants, listeners, viewers, critics, and opponents. Web communication has in this way benefited white supremacist organizations in the United States, for example, by improving organization and recruitment through lowering costs for a geographically scattered group.[59] Because the Web transcends territorial limits in this manner, analysis needs to reconceive 'community' beyond physical proximity and territorial limits. The Web allows specific communities to ignore national borders to a greater extent than ever before. List servers potentially draw participants from around the globe into direct communication over specific issues.[60]

Web communication alters the existing conditions of states and extends messaging capacities for a range of ethnic, religious, racist, and ideological non-state actors. While it circumvents states, the Web is subject to other globalizing or even state-linked influences such as corporations, political ideologies, religion, ethnicity, the environment, other media, and even – as seen in the northeastern United States and Ontario in August 2003 – electrical blackouts.[61] As a result, analyses must account for the influence of offline realities. Stated otherwise, older social forms influence new electronic conditions and produce even more complex elaborate relations: 'each new synthesis generates its own contradiction without necessarily eliminating previous forms.'[62] Global networks have limits since, in many settings, individuals are restricted by states while offline or have no hope of access. IT also gives governments a potent tool to monitor, regulate, and control.

The three-hundred-year dominance of states in international and domestic affairs is not over. Recognizable states clearly survive displacement of certain functions, and their capacity to act, achieve certain ends, and articulate defining narratives is transformed, not terminated. National governments persist, and since governments use 'economic competition as a tool of political strategy, boundaries and cleavages between major economic regions are likely to remain for a long period.'[63] Web activism is a facet of globalization and complements existing dynamics – it is not an independent variable. Non-state actors especially benefit from Web access when conditions for open opposition are not available, and, like all political movements, they recognize the power of communication. Web activist techniques and contents depend on offline political realities. The effect of Web activism is less to overwhelm states than to embody a 'global village' phenomenon that is important for how the world is understood and articulated. However, we need to bear in mind that perceptions and language (including visual ones) do not perfectly coincide. So while the global village points to a sense that electronic communication is a meaningful human change, the specifics of that alteration depend on a range of factors that extend beyond communication. In concrete terms, the Web is another way through which problems in one state, whether humanitarian or political, have become transnational and entered the political arena of other states.[64] The Web is another context in which politics occur, issues are generated, and specific limits pertain. The difference is that the context of concerns and issues has broadened. At a national level, Web activism is paralleled by the emergence of 'e-government,' 'e-services,' and 'e-governance,' all of which suggest the durability and adaptability of states.

The Context of the Global Mediascape

The predictions, hopes, and expectations of dramatic, technologically driven transformation that accompanied the Web's appearance are not unique to our time.[65] Improved printing techniques, rising literacy, and the end of royal censorship also created new media and a receptive public at the time of the French Revolution:

> The first months of the Revolution saw an extraordinary explosion of periodical publications: 42 titles between May and July 1789, more than 250 titles in the second half of 1789. The explosion was prepared, after the meeting of the Assembly of Notables, by multiplication of lampoons, certain of which appeared in many volumes at regular intervals, in this way announcing periodicals. It was facilitated by manufacturing conditions:

material needed to publish a newspaper were not expensive; a single person could write, edit, print and sell a newspaper. The revolutionary paper, written on mediocre paper, crammed with printing errors, is as breathless a manufacture as the reality about which it commented and almost always linked to the personality of a single writer who did not sign articles, but made the entire newspaper.[66]

During the political crises of France's *ancien régime*, pamphlets expressed 'confrontation between defenders of the monarchy and ministerial "reform" on the one side, and the partisans of the *parlements* and some putative historical "constitution" on the other.'[67] The French revolutionaries embraced newspapers and pamphlets as tools to enhance their own and others' participation.[68] They believed that print media would draw individuals into a public sphere, facilitate expression of authentic social interests, introduce 'reason' to political life, and realize harmonious cooperation. Radicals argued that the press 'can teach the same truth at the same moment to millions of men; through the press, they can discuss it without tumult, decide calmly and give their opinion.'[69] Independent newspapers and political pamphlets became widespread media forms[70] that had ambiguous effects. They facilitated debate, expressed socio-political divisions, spread ideas, and reached large numbers of people. However, they did not result in direct democracy or social harmony. As compulsory public education further widened the reading public, the restoration of censorship under Napoleon circumscribed the information and ideas that had been unleashed in the 1790s. The revolutionary press was crushed and the state monopolized print technology. In the mid-term, print technology facilitated control and regulation rather than popular participation.[71]

The example of the French Revolution suggests that technological innovation does not inexorably lead to greater participation. While wealthy societies can produce better-educated citizens who are more assertive about their interests, beliefs, and rights, the well-known and extensively analysed example of Weimar Germany shows that a society that is relatively wealthy and literate can also spawn fascism. While Web activism has spawned interesting interconnections and transgressions in global politics, it is by no means certain that it is a device that will enhance social justice or progressive politics. The example of the Nazis, who used radio to spread their message, is one part of the backdrop to Web activism.

Like the printing press, the Web has sparked optimism and anxieties in a context in which an 'information society' has been constructed as a meta-narrative.[72] In this environment, the Web facilitates information exchange and digitally transposes political 'stories'[73] that displace 'the question of freedom from a presupposition of and a conclusion to theory to become instead a pretheoretical or nonfoundational discursive preference.'[74] Web narratives join a context in which politics already has become 'largely a contest for control of television images.'[75] Identity in the information society is epitomized by television sound bites that electronically construct reality: 'much as image-conscious coverage of political campaigns directs our attention away from the substance toward the packaging and stagecraft, recent accounts of identity emphasize the sense in which we construct, even fabricate ourselves.'[76] Web activism thus aims to inject alternative content into a highly stylized political culture in which cost, corporations, and access substantively marginalize specific issues and conditions. It aims to take advantage of the rise of multiple and overlapping forms of authority in a global-scape of networks. Suggestions of radical political transformation that enhances individual input or creates community should be qualified due to the Web's novelty, limited access, increased regulation *à la* The USA Patriot Act, and persistent countervailing tendencies.[77] Radio and television were also heralded as tools for new individualism and communities before they greatly increased the role of interest groups.[78] The Web's capacity to inject groups from outside the mainstream into the mediascape may be its greatest impact, but the values, issues, and interests that it carries could be diluted, ignored, or distorted along the way. In the short term, the articulation of national and global interests[79] will change since players and ideas

> exist in the power game through and by the media, in the whole variety of an increasingly diverse media system, that includes computer-mediated communication networks. The fact that politics has to be framed in the language of electronically based media has profound consequences on the characteristics, organization, and goals of political processes, political actors, and political institutions.[80]

The disappearance of the MRTA website after 1997 suggests that sustaining non-state access to a wider public is indeed problematic. Web activism flourished due to fragmentation, eroded national boundaries, increased cynicism about governments, and awareness of transnational issues such as the environment and human rights. After 9/11, these conditions were overlaid by the Bush administration's obsessive focus on fighting 'terrorism' and protecting the state. The obsession translated into campaigns centred on states in Afghanistan and Iraq that, in the early 2000s, failed to provide stable conventional governance. Since Web activism co-exists alongside the so-called war on terrorism, it illustrates the paradox of a global-scape that is both state-centred and multicentric.

Limited access in most regions and for most of the population of the world obviously limits the Web's capacity to foster political participation. Infrastructure, funding, computer literacy, network access, educational levels, and economic resources need to be much more equitably distributed than at present to reach this end. Additional limits are also apparent. Online information requires time, energy, and mental effort to sort, filter, interpret, and utilize. Given that fragmentation, regulatory breakdown, cynicism, and greater awareness during the French Revolution were followed by a universalistic, regulating, and uncritical Napoleonic dictatorship, it is not clear 'whether, on balance, communications have served to empower citizens or states'[81] since human realities do not develop in straight lines or according to social science models. The latter are, after all, only intended as thumbnail approximations to aid discussion, debate, and reflection. The results of improved communication above all depend on content since, as Sontag notes, 'media are essentially contentless.'[82] The many fractures, layers, and divisions of global civil society provide content. In some cases, IT empowers citizens (in the Philippines, cell phones were used to oppose the Marcos government, and, in Iran, audiocassettes transmitted illegal information from Ayatollah Khomeini in the 1970s).[83] In other cases, governments benefit. In China, Web technology is used for state development ends, Web activists are arrested, and the mass of the population is not online:

> Even as the so-called wired elite mushrooms and gains influence, growing numbers are arrested for expressing antigovernment views online. Falun Gong followers who use the Internet to spread information are sent to reeducation camps. Meanwhile, millions outside China's urban centers still lack telephones, much less Internet access.[84]

These uneven conditions spurred Jean Baudrillard's argument that IT has 'no other meaning but that of binding humankind to a destiny of cerebral automation and mental underdevelopment.'[85] The Chinese and Saudi governments' firewalls suggest that his vision is not outrageous.

In spite of its limits as a social force, Web activism is one element in a changed situation. James Katz and Ronald Rice suggest that 'although the Internet has not led to any political revolutions, it has supported and encouraged them (as have – and do – the phone and fax).'[86] Few activists and non-state actors want to abolish states.[87] States still wage war, make peace, are the central focus of many political movements, and are key international actors.[88] In a fragmented global-scape, Web activism unsettles specific governments in particular circumstances and substantiates a new global bipolarity 'between abstract, universal instrumentalism, and historically rooted, particularistic identities.'[89] Even this impact might eventually be subsumed by other factors. In the meantime, like the electronic machines that began to mediate discourse in the nineteenth century, the Web changes how interests are expressed: 'machines enable new forms of decentralized dialogue and create new combinations of human-machine assemblages, new individual and collective "voices," "specters," "interactivities" which are the new building blocks of political formations and groupings.'[90] Web activists show how information and communication sustain the guts of political conflict: the will to deliver a message. As such, Web activism is part of a shifting narrative. Continued Web activism depends on social and electronic conditions since, even more than French revolutionary pamphlets, it might disappear without physical trace. Although the complexity of a system that connects computers around the world means that it is extremely difficult to eliminate *all* traces of a specific piece of information, Web communication has a physically fragile existence. Without electricity and systems maintenance, the Web might rapidly fade from human memory.

2 Insurgency Online as Networking: IRSM Web Activism[1]

As the theories of Benedict Anderson, David Miller, and Tom Nairn illustrate, nationalism has spawned a wide-ranging literature. Nationalism does not resemble liberalism, conservatism, fascism, communism, socialism, or the other major political ideologies of the post-1789 political world. Nor does it fit the materialist assessments of both left and right. For much of the twentieth century, nationalism was seen as redundant, a sign of political and cultural underdevelopment that had been relegated to the sidelines of political history by a global capitalist-communist struggle. Both the United States and the USSR were multi-ethnic, multiracial, and multicultural entities in which *difference* and specific identities were allegedly subsumed beneath industrial progress. Countervailing tendencies periodically burst forth as Ukrainian nationalism, black nationalism, or political Islam but were not regarded as signposts on the motorway of human development. The emergence of a multitude of national identities on the global scene since 1989 thus came as a surprise to the theorists and practitioners of global bipolarity, who continue to apply that template in Afghanistan, Iraq, Palestine, Burma, Ukraine, and elsewhere.

In spite of this conceptual negligence, the political forms of nationalism are resilient and adaptive. Benedict Anderson describes them as *imagined political communities*. For Anderson, nationalism is unlike other political ideologies because it always appeals to a specific and limited public rather than making universal claims. The specificity that makes nationalism difficult to understand alongside the 'modernism' of liberalism, conservatism, socialism, communism, and other ideologies is that its appeal is based on place and identity instead of needs. In contrast, ideologies like liberalism, conservatism, and communism assume that the fulfillment of needs will facilitate a sense of place and identity. Anderson says national community is imagined because its members never know the majority of the cohort with which they identify. Imagined communities also have built-in boundaries because 'even the largest of them, encompassing perhaps a billion living human beings, has finite, if elastic, boundaries, beyond which lie other nations.'[2] Despite this, nation is conceived as an arena in which people's freedom can be realized on the basis of identities or socially constructed relations. Nationalists' ability to imagine a fraternity (like other ideologies, nationalism *is* a boys' club) has given the ideology its great reach since the eighteenth century insofar as it has been adopted around the globe.

Nationalism is anomalous in other ways. Tom Nairn, like Anderson, argues that 'the true subject of modern philosophy is nationalism, not industrialisation; the nation, not the steam-engine and the computer.'[3] Nairn says modernization has a fragmenting impact that in fact stimulates a social reaction that focuses on *ethnicity*. The latter is a way to preserve community and self-identity in the face of uncertainty and change. Ethnicity is a narrative intended to give coherence to public life by linking individuals on the basis of language, history, religion, and/or other features. David Miller notes that nationalism has advantages in that

> it provides the wherewithal for a common culture against whose background people can make more individual decisions about how to lead their lives; it provides the setting in which ideas of social justice can be pursued, particularly ideas that require us to treat our individual talents as to some degree a 'common asset,' to use Rawl's phrase; and it helps to foster the mutual understanding and trust that makes democratic citizenship possible.[4]

Miller sees nationalism as a celebration of an attachment to a historic community.[5] By casting nationalism as imagination-narrative, modernism, and democracy, Anderson, Nairn, and Miller provide a way to understand the phenomenon in its many contexts.

The narrative and identarian features of nationalism are reproduced in e-nationalism, which shares the former's basis by centring individuality and community in stories about groups of people. Both nationalism and e-nationalism are narratives that outline the features of a particular group. Unlike nationalism, which is strongly coloured by historical attachment, e-nationalism is a different type of political phe-nomenon because history is only part of its make-up. As Nairn observed regarding nationalism, e-nationalism is also a type of response to modernization. E-nationalism is about *representation* of a place as the home to a specific group of people, but it is not directly tied to a physical space or territory. E-nationalism is instead tied to rep-resentation of a network of relations based on a common language, his-torical experience, religion, and/or culture. As illustrated by the case of akakurdistan.com, e-nationalism is about both memory and the future projection of a place as *the* home for a given group. The articula-tion or representation of e-nationalism is transnational in character, which contrasts with the spatial focus of traditional nationalism.

Nationalism sometimes includes an appeal to a transnational diaspora.[6] In e-nationalism, the diaspora is linked to the mediascape and so becomes an *active agent* of the movement instead of an *episodic participant*. E-nationalism is different in another important way. Fou-cault points out that the end of nationalism is the development and knowledge of a state's power in competition with another state.[7] E-nationalism reconfigures competition out of territory and into the mediascape. Since the search for a balance of territory led to the state-based international order, it is possible that the search for an equilib-rium of skills, infrastructure, and knowledge could produce a new form of global power. All three cases in this book show the anti-regime or insurgent aspect of that power. The IRSM, RAWA, and MRTA web-sites illustrate a form of anti-regime power in distinct ways through: global witnessing (RAWA), serving as media relay (MRTA), or net-working (IRSM). Together, the three cases portray a variable set of political behaviours. Some observers note that the Web has

> led to the concerted development of politicized groups that use both the Internet and the Web to inform, connect and coordinate interests beyond geographic boundaries. The anti-globalization movement has been partic-ularly successful at linking disparate groups into unified movements of resistance ... the new flows of informational news sources via the Web explain how new forms of oppositions can develop and prosper through wider rings of interconnection. The very concept of community is being redefined through the use made of informational new possibilities and potentials.[8]

E-nationalism could develop in several directions that follow the general framework of David Held's globalization thesis. On the one hand, Web nationalists could adopt a sceptical stance regarding globalization and present e-nationalism as a response. Alternatively, Web nationalists could become 'transformationists,'[9] who seek change through a globalization process in which, they realize, 'shared membership in a political community, or spatial proximity, is not regarded as a sufficient source of moral privilege.'[10]

E-nationalism and the Electronic Starry Plough: Insurgency as IRSM Web Activism

The Starry Plough[11] is the emblem of the Irish Republican Socialist Movement (IRSM). The image was first used as the official symbol of the Irish Citizen Army (ICA) in 1914. The early version of the Starry Plough was set on a background of green and bordered with a gilt fringe. The flag flew above the Imperial Hotel in Dublin during the 1916 rising. The IRSM adopted and modified the flag, placing white stars on a blue background. The organization uses the image to point to its links with Ireland's struggling class warrior predecessors, militant Irish socialism and ICA leader John Connolly in particular. As such, the Starry Plough is a symbol of working-class militancy as well as a flag of rebellion, class war, and social revolution. The electronic Starry Plough in the title of this section refers to the transposition of the IRSM symbol into the mediascape of the Web.

Social and political movements base their appeals for support and messages to the public in identarian terms. They use symbols to define the lines of debate and establish boundaries. Boundaries can be moral, as in the case of *charivari*, which was used in medieval Europe to shame, taunt, and intimidate, or have more abstract purposes, enclosing 'elements which may, for certain purposes and in certain respects, be considered to be more like each other than they are different.'[12] Conventional political boundaries have been territorial since the seventeenth century. The ways in which conventional political boundaries have been expressed include flags, parades, and national symbols. Our experience of political boundaries also includes identities and political societies that do not strictly match physical limits. An example of this is 'Irishness,' which has specific meanings for the inhabitants of the island but also resonates in an effective sense (as motivation, value system, and structure for beliefs) for people in the UK, North America, Australia, and elsewhere. This symbolic representation of politics is also reproduced among non-state actors in politics. The IRSM is an example of one such group, which has now introduced its symbolism into the global mediascape through Web activism.

Northern Ireland and Ireland: A Mediascape for the IRSM

The IRSM context is particularly complex given that the focus of its concerns and organization straddles two distinct sovereign states, the UK and the Irish Republic.[13] A key IRSM concentration is Northern Ireland and the thirty-year civil conflict known as 'the Troubles.' The conflict grew on the basis of divisions between Northern Ireland's 1.69 million inhabitants, who are split between Catholics (38.4 per cent) and Protestants (50.6 per cent).[14] The conflict has been exacerbated by Northern Ireland's economic decline over the past century. Civil conflict and economic decline produced a distinct set of social conditions: at the end of the 1980s, about 40 per cent of the workforce was employed in the public sector, especially in security-related areas such as policing and prisons. The region has recently experienced significant growth,[15] but Northern Ireland developed a dual economy, with people employed in the public sector leading relatively comfortable lives, contrasted with those who are unemployed, poorly paid, or working part-time.[16]

The dual economy is characterized by identity-based inequalities. In 1991, Catholic males were over two times more likely to be unemployed than their Protestant peers. Inequality is also present in terms of occupations. In 1991, 29 per cent of Catholic males were white-collar workers as opposed to 39 per cent of Protestant males. Over the last thirty years, Protestant males have been over-represented in security-related occupations and services, management and administration, and skilled engineering, while Catholic males have been over-represented in lower-level manufacturing and construction jobs. A historically disadvantaged Catholic community has made gains over the past thirty years, but its overall socioeconomic position is still inferior to that of Protestants. The Catholic middle class has moved beyond serving its own community to assume positions of responsibility in the public sector but still has not achieved proportionate representation in occupations and economic equality with Protestants.[17]

Politically, the province or 'statelet' of Northern Ireland has undergone many mutations. Territorially, it incorporates six of the nine counties in the historical province of Ulster. Because it does not incorporate the entire province, the IRSM views it as a political misnomer to call the region 'Ulster' (as Protestant unionists and loyalists do) and instead refers to it as Northern Ireland. The region's political system is based on principles of representative democracy and individual rights and liberties that are regularly expressed through free and contested elections based in universal suffrage. However, the practical functioning of liberal democracy is not clear-cut[18] because political structures have been subject to constant change and protest. From 1921 (the date of partition and creation of the Free State in the south) to 1972, Northern Ireland was governed by a regional parliament at Stormont. Ironically, this Home Rule was opposed by Unionists when they were part of a single Ireland. The Stormont assembly legislated local affairs and left foreign affairs, trade, taxation, and defence to the Westminster Parliament. Stormont was abolished by the UK parliament in 1972, after years of civil unrest and political violence. The British government then assumed full control of Northern Ireland from Westminster in an arrangement called 'Direct Rule.' From 1974 until the Good Friday Agreement was implemented in 1998, the Westminster parliament approved all laws for Northern Ireland and a secretary of state controlled local government departments.

In the 1970s and 1980s, the British government attempted to devise a settlement to reinstate regional government, but failed due to the intransigence of Unionists and a boycott by nationalists. The Good Friday Agreement (GFA) of 10 April 1998 is the most recent attempt at establishing a constitutional settlement. The GFA is based in community safeguards and a 108-seat assembly that is proportionately elected by single transferable vote. The new assembly, which has been repeatedly suspended since its inception because of political infighting, has legislative and executive authority over agriculture, economic development, education, the environment, finance and personnel, and health and social services. A nominating process for candidates to the executive is designed to provide equal representation of the nationalist-republican and unionist-loyalist parties.

Electoral manipulation has been the greatest historical obstacle to a properly functioning liberal democracy in Northern Ireland. Given the long-standing antagonism between the Catholic minority and Protestant majority, voting patterns have split along ethnic lines. The electoral system is in fact one of the main ways in which different identities are expressed. Protestants mainly vote for unionist parties that advocate maintaining the Union with the UK and preventing the assimilation of Protestants by what is seen as a monolithic Catholic Ireland. The evolution of the electoral party system has also played a role in the history of the Troubles. In the period immediately after partition and creation of the Free State in 1921, the Northern Ireland electoral system was based on a proportional representation (PR) system of single transferable vote (STV). By the mid-1920s, PR led to the emergence of political parties that threatened the ruling Ulster Unionist Party (UUP). The electoral threat came from nationalist and republican organizations, marginal Protestant parties that wanted to break ties with Britain and form an Independent Northern Irish State as well as labour and socialist parties.

Realizing 'that PR had allowed the electorate a measure of choice incommensurate with Unionist Party ambitions,'[19] the UUP government replaced PR with the British plurality or 'first-past-the-post' system that polarized elections between unionism and nationalism. Plurality voting drastically impaired electoral competition: Unionists represented 66 per cent of the population but controlled 85 per cent of all local authorities by the late 1920s.[20] Given the high Catholic birth rate, the unionist community feared that its regional and local government power would evaporate and so manipulated elections by redrawing constituency boundaries. PR was reintroduced in 1973. Abolition of the plurality system started to fragment the Unionist vote and strengthened the electoral power of the Social Democratic and Labour Party (SDLP) and Sinn Fein. However, PR only applied locally and Direct Rule meant that real political power was held by the Westminster parliament, elections to which were determined by plurality voting. Slicing electoral boundaries through communities, which created Catholic electoral minorities out of demographic majorities, perpetuated the high degree of political polarization in Northern Ireland.

Another barrier to democratic rights and freedoms in Northern Ireland developed from the way in which the UK government responded to the Troubles. The Prevention of Terrorism Act (PTA) was implemented as a temporary measure in 1974 in response to authorities' need to 'fight against terrorism.' One section of the PTA allows authorities to detain suspected terrorists for as long as seven days without charge. Critics of the PTA argue that this section breaches the European Convention on Human Rights and infringes on basic civil rights in a liberal democratic state. The PTA also prevents suspected terrorists from entering other parts of the Union (i.e., Scotland, England, and

Wales), limiting individual freedom of movement within a single sovereign state.

Threats to freedom of speech also exist. Although many newspapers either implicitly or explicitly express sectarian sentiments, some observers 'suggest that sections of the media have served the interests of the British state by acting as channels for the dissemination of government propaganda.'[21] The state has periodically placed real limits on freedom of speech. Between 1969 and 1993, about one hundred television programs on the Troubles were banned in the UK. In 1988, the UK government imposed a ban on all programs that contained interviews with individuals either directly or indirectly linked to illegal paramilitary organizations. The ban was not revoked until 1994. Beyond legal measures, the media have also been affected by self-censorship due to the threat of legal sanctions and restrictions on media production.[22]

The Republic of Ireland is the parallel context in which the IRSM is active. It presents dramatically different conditions from the North. The Republic has recently had extremely strong economic growth and experienced profound social, economic, political, and cultural transformation since the 1980s. The economic growth rate in the 1990s was about 8 per cent per year, and unemployment declined from 18 per cent in 1987 to 4.6 per cent in 2003.[23] In 1999, the GDP increased by an estimated 9.5 per cent, and incomes have started to increase for all groups in spite of inflation.[24] In the early years of the new millennium, Ireland's growth and transformation continued. The economy has been transformed from 'being agrarian and traditional manufacturing based to one increasingly based on the hi-tech and internationally traded services sector. In 2003, the services sector accounted for 66% of employment, industry for 28%, and agriculture for 6%.'[25]

The Republic is a parliamentary democracy with an electoral system based on proportional representation in which political issues now centre on managing the consequences of growth: for example, income inequalities, corruption, and new social values. As Kearney and others suggest, the overall picture is a dramatic transformation of an impoverished and highly religious society on the margins of Western Europe into a dynamic social, economic, and cultural formation. Once at odds with Western European norms, the Republic is now closer to the mainstream. Religious observance is higher than in most advanced industrial societies, but steadily declining. Reflecting these conditions, the IRSM focus on the Republic emphasizes social issues rather than nationalism.

Website Analysis – http://www.irsm.org/irsm.html

The analysis of the IRSM website was conducted in June and July 2000. Additional materials since added to the site are not examined here. All references to the IRSM website are to the June-July 2000 version, which is used as a 'snapshot.' In 2000, the IRSM provided an umbrella for a number of organizations that are included on the website: the Irish Republican Socialist Party (IRSP), the Irish National Liberation Army (INLA), the British wing of the IRSP, and the Irish Republican Socialist Committee North America (IRSCNA). The website[26] was divided into sections on IRSM history and principles, group statements and communiqués, the IRSM 'Honour Roll,' the IRSP, IRSCNA, IRSP London, contact information, and links. It was also organized according to kinds of documents. In May-June 2000, the site featured approximately 267 texts, ranging from less than a page to ten pages in length. The shortest and most numerous texts were political communiqués and the longest were historical pamphlets. Except for press releases, all documents on the website are undated and anonymous.

The first group of documents on the site are analytical treatments of the domestic situation in Northern Ireland and especially the Troubles. The IRSM explains its interpretation of the Troubles and proposes ways to resolve the conflict. Its analysis includes theoretical texts on Ireland and the struggle for Irish national liberation as well as historical documents by Irish republican socialists such as James Connolly, the Irish socialist leader executed for his leadership of the 1916 Easter Uprising. The theoretical texts also examine capitalism and economic development from a Marxist-Leninist perspective. In addition, the site outlines IRSM positions and policies in contrast to other nationalist groups. It also contains a number of statements on a wide variety of topics, with particular emphasis on how IRSM activities respond to domestic and international events. Lastly, several sections on the site commemorate members and support political prisoners. These pages list IRSM members killed in action (the group uses the term 'assassinated'). A 'prisoner of war' section lists IRSM members who are imprisoned as a result of their activities.

IRSM Political Goals and Analysis

The website addresses a number of conflicts. Above all, it focuses on the conflict over British sovereignty in Northern Ireland, which is called a struggle for national liberation. The site also features the IRSM struggle against capitalism and outlines the organization's goal of creating a workers' republic in Ireland. In addition, the site addresses the conflict between the IRSP-INLA and other Republican movements, particularly Sinn Fein (SF) and the Provisional Irish Republican Army (PIRA) as well as the Official Irish Republican Army (OIRA). The IRSP-INLA carefully distinguishes its republican socialism from the republican nationalism of SF-PIRA by advocating national liberation of Northern Ireland and establishment of a workers' state on the entire island.[27] The movement distinguishes republican socialist national liberation struggle from mainstream nationalism, characterizing the latter as a form of chauvinism produced by advanced capitalist states. The IRSM charges that Britain has an imperial ethos and foreign policy while contending that national liberation is a struggle to free national territory by expelling external, imperial powers. In this way, the IRSM identifies its struggle in Northern Ireland with others in Cuba, Vietnam, Angola, and Algeria. The IRSM defines itself as a socialist organization that is committed to armed struggle. It says this stance sets it apart from republican organizations that have abandoned class struggle and define themselves as mainly nationalist, such as SF-PIRA. The IRSM alleges that SF-PIRA policies conform to the material interests of the Catholic petite bourgeoisie. At the same time, the IRSM distinguishes itself from what it calls the 'reformist' socialism of the OIRA (now the Workers' Party) from which its organization emerged.[28]

The IRSM says it is part of Ireland's 'green' Marxist history. Green Marxism is based on the writings and activism of James Connolly.[29] He argued that British imperialist policies were the 'prime mover' of the 'Irish Question' and that the causes of conflict were exogenous. According to Connolly, after British imperialism seized Irish Catholic land in the North and settled Scottish colonists on the best acreage, the UK split the Protestant and Catholic working classes through a 'divide and conquer' colonial policy. He thought that the only way to solve Irish problems was to get rid of the British presence by a national liberation war and the establishment of socialism. National liberation and socialism were inextricably linked for him. IRSM Green Marxism follows Connolly's theory, arguing that the Troubles result from British imperialism.[30] The IRSM argues that the British used partition to divide the Irish working class between Catholic republicanism and Protestant loyalism. In Northern Ireland, unionist parties used the cleavage to manipulate legitimate Protestant working-class concerns over civil and religious liberties in a united Ireland and tie its material interests to British imperialism. The IRSM says the Catholic working class was duped by conservative republicans into believing that its interests were identical to those of the Catholic middle class.[31]

The IRSM views the governments of both Northern Ireland and the Republic as illegitimate and positions itself as anti-regime insurgents. The IRSM sees the North as an imperialist statelet set up by a foreign occupation force and, as such, impossible to reform. The organization sees the only solution to the conflict as reunification of North and the Republic and their transformation into a socialist workers' state. The Irish Republic is seen as an illegitimate 'bourgeois class-state' that gave up armed struggle against British imperialism. The Republic is taken to task for persecuting militant republicanism (particularly socialists) and 'watering down' its claim to sovereignty over the six counties of Northern Ireland. It is viewed as a state that is beholden to the 'medieval' interests of the Catholic Church, which further delegitimizes it in IRSM eyes.[32]

IRSM Revolutionary Principles and Policies

As a revolutionary party with Marxist-Leninist principles, the IRSM believes in the dialectical unity of revolutionary theory and practice. This means that the organization aims to put revolutionary ideas into action. For the IRSM, revolution would replace the social, political, and economic system for all of Ireland with an entirely new order based on what it sees as people's needs, their welfare, and national independence.[33] The IRSM revolution is a two-phase process based on (1) an end to partition and seizure of the state and (2) the setup of a 'dictatorship of the proletariat' and construction of socialism under IRSM leadership. The IRSM would abolish parliamentary democracy and the party system and replace them with a system of representation built around a single revolutionary party.[34] The IRSM distinguishes the principles in its revolutionary program from its policies, which are determined by given historical situations and thus more flexible. The organization views policies as a matter of strategy, and tactics and principles as goals.

IRSM policies are divided into two categories: (1) power policies and (2) reconstruction policies. Power policies are geared toward the revolutionary seizure of state power; reconstruction policies are designed for a subsequent phase in which socialism would be built. The IRSM sees armed struggle as a power policy (i.e., not a matter of principle) that can be renounced if it furthers the organization's revolutionary principles.[35] As a political party, the IRSP has formulated a series of positions:

- Britain's renunciation of its claims to sovereignty over Northern Ireland or any part of Ireland;
- dissolution of the Ulster Defence Regiment (UDR), Royal Ulster Constabulary (RUC), and Royal Ulster Constabulary Reserve (RUC Reserve), and withdrawal of British troops from Ireland;
- release of all political prisoners and the granting of amnesty;
- British compensation for those who suffered from imperialist violence and exploitation;
- opposition to imperialist control over Ireland's resources and wealth;
- rejection of partition and recognition that both states are illegitimate;
- convening an All-Ireland Constitutional Conference to develop a democratic and secular constitution;
- armed struggle as an inherent right of the Irish people to achieve self-determination and national liberation.[36]

Beyond revolution, the IRSM articulates a point of view on a series of issues in everyday life in Northern Ireland. The party argues that the British-led sectarian state in Northern Ireland deliberately fosters working-class sectarianism as well as the division of the women's movement between support for national liberation and espousal of pro-imperialism.[37] It charges that the dominant role of the Church in Irish society strengthens patriarchy because it perpetuates social values that oppose abortion, divorce, birth control, and women's entry to the workforce.[38] The IRSP argues that the power of the Catholic Church in society must accordingly be limited. The group notes that while birth control is legal in the Republic, those using it are stigmatized, Catholic doctors often refuse to provide contraceptives, and women need their spouses' consent to be sterilized, which implies that they belong to men. The IRSP advocates easy and unconditional access to birth control. Given that abortion is for all intents and purposes illegal in the Republic and North, the IRSP supports pro-choice movements and legalized abortion. It also supports the right to divorce, pay equity, free child care, curtailment of health care cutbacks, tougher sentences for rapists, an end to strip-searching, and a halt to discrimination against lesbians.

Communiqués

By 2000, the IRSM had released about 129 statements on domestic Northern Ireland issues[39] that covered areas such as: Loyalist violence; the peace process; UK troops and the British government; the RUC and police reforms; prisoners of war; the Derry Monument to hunger strikers; domestic calls of solidarity; unionism and parading; and miscellaneous political statements. The contents of statements range from commentary on policing policy in Northern Ireland to condemnations of British or Irish government positions on pay equity to 'revolutionary statements' released for May Day that reaffirm the necessity of building socialism and fighting for workers' rights to denunciations of loyalist violence.

The Northern Ireland peace process is often addressed. The IRSM opposed the Good Friday Agreement (GFA) and termed it a failure. At the same time, the organization supported SF in the GFA election and called for no compromise in negotiations over arms decommissioning, a process in which it played no role. The INLA provisionally renounced armed struggle while the IRSM condemned British security forces and the RUC for violence and corruption. A final group of statements on the website focus on Unionist actions, parades, solidarity with Irish lesbians and gays, and anniversary dates in the armed struggle. Many IRSM communiqués respond to foreign events, either as memorials and/or commemorations,[40] calls of solidarity,[41] or condemnations.[42] INLA communiqués largely focus on armed struggle. Most commemorate volunteers who died for the cause while others take responsibility for attacking military targets, deny responsibility for assassinating loyalists, or address cease-fires. Website links cover a wide range of movements and organizations in categories that include republican socialists,[43] solidarity links,[44] other pages of interest,[45] and other documents.[46]

Conclusions

The electronic Starry Plough suggests how a shift from territorial politics to a system of multiple and overlapping interests, meanings, and authorities might be underway due to the varied forces of the globalscape. The targets and impacts of IRSM Web activism cannot be measured in relation to a specific territorial unit as has traditionally been the case under the conventional model of political societies since 1789.

Instead, the IRSM focuses on issues in a mediascape (though *not* a territorial unit since it concentrates on both the Republic and Northern Ireland) and addresses a geographically dispersed constituency. The constituency is spread over at least three continents that are distant from one another (Australia, North America, and Europe). A system of overlapping and multiple meanings, interests, and authorities did exist prior to the contemporary mediascape, but not in the same manner. The advent of Web activism facilitated cohesiveness, communicativeness, and an ability to establish a functioning network for a community of support that it would not have been possible without globalizing IT. The effect is significant for a marginal non-state actor like the IRSM, which can now systematically assemble members, foster discussion, and spread its core principles and ideas.

The Web is a key element in IRSM's transformation from an ideologically and socially marginalized organization on the fringes of Europe to a functioning global network. This transformation embodies how IT-driven skills, knowledge, and infrastructure have influenced IR. The transnationalization of IRSM activism has been paralleled by an internationalization of the process of conflict resolution in Northern Ireland, seen in U.S. government involvement in the GFA under the Clinton administration and the creation of the arms decommissioning body for the region. Rosenau argues that such IT-linked change has profound implications for global politics:

> With people in both developed and developing countries becoming more skilful in relating to public affairs, with organizations proliferating at an eye-catching and accelerating rate, it is hardly surprising that information technologies have contributed to transformations in historic global structures. Stated most succinctly, as the global arena has become ever more dense with actors and networks, the traditional world of anarchical states has been supplemented by a second world of world politics comprised of a wide variety of nongovernmental, transnational, and subnational actors, from the multinational corporation to the ethnic minority, from the professional society to the epistemic community, from the advocacy network to the humanitarian organization, from the drug cartel to the terrorist group, from the local government to the regional association, and so on across the whole range of collective endeavour ... this interaction between the worlds has been facilitated and intensified by the information technologies, thus collapsing time, deterritorializing space, and rendering traditional boundaries increasingly obsolete. Indeed, the more the tech-

nologies advance, the more they have facilitated the opening up of both governments and nongovernmental organizations to the influence of their members, to bottom-up and horizontal processes that have greatly complicated the tasks of governance on a global scale.[47]

Web activism alone will not radically transform politics to serve individual input or create community. As stated above, radio and television also initially seemed to herald a new individualism and new communities. Their eventual impact in the form of a greatly enhanced role for interest groups was not foreseen when the technologies were first introduced. To have impact through the Web, a group must either attract a broad public or mobilize a particular constituency. Irish republicanism fits this global mediascape since it has a long history of successfully mobilizing its constituency. Republican efforts have ironically been aided by the de-territorializing (and globalizing) impact of events such as the tragic nineteenth-century famine, which eventually led to an affluent and sympathetic Irish-American relay for the republican movement. The IRSM has used the Web to spread its much more marginal interpretation of republicanism to the diaspora. The Web's availability to non-state actors outside the mainstream provides them with opportunities to spread messages that Manuel Castells says will have 'profound consequences on the characteristics, organization, and goals of political processes, political actors, and political institutions.'[48] Before the advent of the contemporary mediascape, these messages would have spread more slowly or not at all.

Irish republicanism has a long history of successfully attracting a public, even on the back of globalizing tragedies such as nineteenth-century famine. In terms of a specific constituency for the IRSM, the Web provides groups outside the mainstream of politics with an opportunity to spread their message. As such, the Web challenges assessments of non-state actors' IT use since the first task 'is to understand the genuine complexity of this new medley of technologies and to anticipate how decisions and policies concerning its potential benefits and dangers could affect its future values to society.'[49] My reading of the IRSM website illustrates how a non-state actor used IT to network and propel a marginal variant of Irish nationalism into the global mediascape rather than basing itself in hyperbolic claims about 'hypertext' or 'networks' as does some IR research.

IRSM Web activism benefits from today's fragmented public, eroding national boundaries, increased cynicism about governments, and widespread awareness of the broad range of transnational issues. However, as stated above, these conditions do not necessarily signal long-term transformation in the direction of greater participation. Web access has grown but has not directly reached most of the world's population. For the foreseeable future, the direct impact of the Web is in economically and culturally privileged areas of the globe. Web activism requires time, energy, and intellectual resources. Information must be sorted, filtered, interpreted, and utilized before it has an impact. As a result, it is not clear 'whether, on balance, communications ... empower citizens or states.'[50] In effect, global communications suggest that we rethink how we understand information. Since simple production or reproduction of information on the Web is not a sufficient impetus to change, its impact and potential might lie elsewhere.[51]

The IRSM shows the centrality of information and communication in political networking in the age of globalization. If Web activism provokes something less than the revolutionary overthrow of hierarchies, it is part of the transformed conditions of sovereignty. Few online insurgent and radical non-state actors want to abolish states. The IRSM wants to transform and extend the territory of a state rather than overturn state-based power. IRSM Web activism unsettles the Irish and British governments and embodies the global bipolarity between so-called universalizing technologies and specific identities.[52] The group's Web activism de-territorializes messages rather than politics. De-territorialized messages might influence de-territorialized politics, but Irish republican political goals are still very much influenced by national divisions set in territorial units. The delivery of messages and a narrative depends on a social context and global mediascape that give meaning. The Irish social context lends significance to an IRSM electronic voice that might otherwise disappear without trace. The IRSM uses the Web to spread a message that is otherwise marginal to Irish, Northern Ireland, and global power. The IRSM message would be unheard if the Web did not exist to breathe life into it.

3 Insurgency Online as Global Witnessing: The Web Activism of RAWA[1]

The marginalization of the concept of gender in discussions of politics and power is surprisingly persistent. Introductory politics textbooks often either ignore feminism or relegate it to a 'special issues' section. Schools of political science that range from libertarians to Straussians (followers of the philosopher, not the composer) explicitly dismiss gender issues as pandering to 'political correctness' or special interest groups. Imagine the surprise and consternation of savvy undergrads who come across Aristotle's assertion that 'it is a mistake to believe that the "statesman" is the same as the monarch of a kingdom, or the manager of a household, or the master of a number of slaves'[2] as if the three categories were the same male prerogative. In conventional political science, women's rights are largely treated as a private matter. End of discussion. At the same time, conventional political science rejects Aristotle's views on slavery:

> it is also clear that there are cases where such as distinction [of the natural slave and the natural freeman] exists, and that here it is beneficial and just that the former should actually be a slave and the latter a master.[3]

This does not refer to S/M sex or Hegelian myths, but actual historic variety of slavery. Political science gender myopia is especially puzzling since the social value of slavery is obviously unacceptable while the validity of gender issues is, in the age of war rape and sex tourism, somehow up for debate in some circles.

Gender-myopic readings of politics thrive on poor historical knowledge of post-1850 social development and the conscious social construction of gender divisions before the First World War in the West.

Such political science is proud of its rigour and historical knowledge but ignores the constructed categories of late-nineteenth-century sexologist Richard von Krafft-Ebing, who

> repeatedly refers to males and females as 'opposite' – anatomical, genital differences signify an all-encompassing fundamental contrariety ... Human males and females are not just different in some biological structures and functions, similar in others, depending on one's standards of evaluation. This doctor's two sexes are antithetical.[4]

Implementing theories of biological-sex difference relegated 'women's work' (and 'women's issues') to the household and private life, where Aristotle long ago placed it.[5] It led to a cultural bias toward female physicality seen in a cult of passive femininity versus assertive masculinity.[6] Western industrial societies with cults of assertive masculinity repeatedly enter intense great-power economic, military, and political competition that culminates in war.

In the period 1914 to 1989, gender issues were largely ignored. Women appeared in international politics as recipients of aggressive masculinity and participants in conflicts.[7] Women sent sons to war, laboured in jobs that male members of families left behind when departing for war, and worked as nurses in the field of battle. However, women were

> normally excluded from all formalized peace processes, including negotiations, the formulation of peace accords and reconstruction plans. Even where women and children were actively involved in sustaining and rebuilding local economies and communities throughout the conflict they are frequently pushed into the background when formal peace negotiations begin.[8]

The persistent invisibility of gender issues in IR is based in a *segregationist* view of economic, political, and social life in which the contributions of men are more valued and politically relevant. In the real world, women play an economic role through 'unpaid work in reproduction, including the maintenance and refurbishment of the current labour force and the raising of the next one, as well as many women's subsistence, farm and informal-sector work, and their unpaid or under-paid community and service work.'[9] Yet global politics obscures women's work. Western women face workplace 'glass ceilings' or give care to

men in employment situations of nominal equality. Women in postcolonial societies such as Afghanistan's Taliban, moreover, face idiosyncratic and oppressive interpretations of 'tradition' that ostensibly
protect them from 'Western immorality.' In such contexts, 'violence is a
part of the domestication of women, whereby their subordination and
service comes to seem natural, so guaranteeing men's access to
women's bodies and to their labour.'[10] In Afghanistan, the issue was
not securing gender equality but regaining women's right to exist in
public. The Taliban was frighteningly successful in pushing women out
of participation in public life and forcing them to remain men's chattel.

Articulating Afghan gender oppression and domination from the
privileged position of a white male in the West is problematic. For one
thing, the variability of non-Western women's lives means that

> Third-World feminisms do not have the luxury of predictability; and a
> feminist theory that would be global in its compass, as in its intentions,
> must expect to be surprised by the strategies, appearance, and forms of
> feminism that emerge and are effective in Third-World contexts.[11]

The task of understanding is complicated by postcolonial realities in
which Afghan feminism is 'doubly other': as a radically disadvantaged
and disenfranchised group in a stigmatized and exoticized non-Western society. This doubly other feminism is post-territorial, has no state
relay, and no institutions and values support it in a given physical
space. Women's human rights do not exist:

> To disobey the Taliban was to die. Soon after the takeover, a group of
> women in the city of Herat marched in protest. According to eyewit-
> nesses, the Taliban surrounded the women, seized the leader, doused her
> in kerosene and burned her alive. Women were sprayed with acid, beaten
> with twisted wires and shot for crimes such as showing their ankles, let-
> ting a hand slip from under the burqa while paying for food or allowing
> their children to play with toys. As for being outside with any man who
> was not a relative, the sentence was death by stoning.[12]

Under the Taliban, Afghan feminists had no social relay, few domestic allies or access, little cash, and no conventional media. The West
turned its back on Afghan feminists and ignored Afghanistan, wring-
ing its hands as irreplaceable statues of Buddha were dynamited, bin
Laden was harboured, Hindus were forced to wear distinctive

badges, aid workers were arrested for 'propagating Christianity,' and women were reduced to lumps of filthy fabric hunched on roadsides, risking their lives by begging for money to feed their children. Even after the overthrow of the Taliban, women's conditions remain daunting:

> the euphoria that followed the Taliban's ouster has worn off. President Hamid Karzai's government has made a point of hiring women at all levels – from street cleaners to cabinet ministers – but women remain heavily underrepresented throughout the work force.[13]

The West turned its back on Afghanistan because the Taliban and its context were not easy to understand or manage.[14] Until 11 September 2001, the Taliban was an irritant on the fringe of the global-scape. The Taliban destabilized the central Asian land mass so dear to Western strategists since the late nineteenth century. The United States was silent over Saudi and Pakistani recognition of and assistance to an oppressive regime in the 'heartland' region that reverberates in Western and especially American strategic thought.

In 1904, Sir Halford John Mackinder identified

> Central Asia as the pivot area of history from which horsemen have dominated Asian and European history because of their superior mobility. With the age of maritime exploration from 1492, however, we enter the Columbian era, when the balance of power swung decisively to the coastal powers, notably Britain. Mackinder now [1904] considered this era to be coming to an end. In the 'post-Columbian' era, new transport technology, particularly the railways, would redress the balance in favour of land-based power, and the pivot area would reassert itself. The pivot area was defined in terms of a zone not accessible to sea power and surrounded by an inner crescent in mainland Europe and Asia and an outer crescent in the islands and continents beyond Eurasia.[15]

Mackinder's shadow over central Asia endures today as gender oppression, global inequities, and political violence draw the region into U.S. foreign policy calculations. In this respect, the diminished Russia and surging China near Afghanistan are great power contenders and pivotal for strategic global power. Whether they gravitate into Western orbit through freer markets or military alliance, Russian or Chinese influence in central Asia has global strategic impact. Regional oil resources in the region and U.S. anxiety over bin Laden even before 9/11 fuelled assessments of central Asia as key to global security. Unable to control central Asia and Afghanistan in the late 1990s, the West destabilized the region by isolating Iran, investing in former Soviet republics, encouraging Turkish ambitions, allowing instability to fester in Iraq and Pakistan, and turning Afghanistan into a global wasteland of religious fanatics, drug dealers, and misogyny.[16]

In Afghanistan, religion co-exists alongside a broad selection of marketable drugs (especially opium). The losers are the mass of Afghan people, who neither benefit from the drug trade nor find temperal succour in religious extremism. They eke out a living. In a tragic twenty-year downward spiral, Afghanistan became a fulcrum of contemporary global issues: gender oppression, land mines, drugs, displacement, ethnicization of conflict, terrorism, religious radicalism, child abuse, prostitution, and environmental ravage. The winners were those who advise, conceive, activate, and implement great-power geopolitics, whether in the arms trade, drug trade, or the oil industry, all of which benefited from regional instability.

Web activist insurgency against the Taliban occurred against a backdrop of complacent, cynical great powers. The Revolutionary Association of the Women of Afghanistan (RAWA) website illustrates how the conditions for global conflict were altered by the fragmentation of the bipolar international system, the rise of new IT, and the emergence of a complex mediascape. RAWA's website shows how IT mobilizes global civil society and transgresses the territorial rigidities of conventional political thought and practice. While conflict once occurred in physical limits based on a political version of the retail adage 'location, location, location,' RAWA turned to a Web-based struggle over perceptions that aimed to use information to shape opinions among international public and global/national decision-makers. Similar trends were seen on 9/11, when innovative tactics and weapons unexpectedly exposed the myth of American invincibility. Ironically, at the same time as 9/11 thrust global violence into the U.S. homeland, RAWA Web activism challenged the Taliban's autarkic vision. RAWA did not win a clear-cut political victory for Afghan women so much as affirm their place on the global agenda by documenting abuse. This place was demonstrated when U.S. president Bush signed the Afghan Women and Children Relief Act of 2001, declaring that 'the people of Afghanistan have suffered under one of the most brutal regimes in modern history; a regime allied with terrorists and a regime at war with women.'[17]

Afghan women's rights were again emphasized by a U.S. State Department spokesperson who stated that

> respect for women is a non-negotiable demand of human dignity. In other words, it does not depend on culture, or religion, or geography. Respect for women is something that the United States stands up for and demands and defends around the world.[18]

The increased prominence of women's issues in the global agenda was also seen in UN Security Council Resolution 1325 (2000), which expressed 'concern that civilians, particularly women and children, account for the vast majority of those adversely affected by armed conflict' and reaffirmed 'the important role of women in the prevention and resolution of conflicts.'[19] By arousing concern, Web activism might have helped prevent an even more catastrophic turn in Afghanistan's still-daunting conditions.

Afghanistan: A Global Wasteland of Drugs, Religion, and Misogyny

Afghanistan's ethnic, religious, and linguistic complexity and agricultural economy lingered in relative geo-strategic tranquillity before 1979. Afghans are divided into four main ethnic groups (44 per cent Pashtun, 25 per cent Tajik, 10 per cent Hazara, 8 per cent Uzbek), while 13 per cent of the population consists of smaller ethnic groups (Aimaks, Turkmen, Baloch, and others). Ethnic diversity is matched by a multiplicity of languages: 50 per cent Afghan Persian (Dari), 35 per cent Pashtu, 11 per cent Turkic languages (mainly Uzbek and Turkmen), and 4 per cent another thirty minor (mainly Balochi and Pashai) languages. In terms of religion, the population was mainly Sunni Muslim (84 per cent) with an 11 per cent Shia minority.[20] Gender-based conflict emerged after post–Second World War social transformations, when urban women in particular were educated, moved into professions, and gained some autonomy. Language, ethnicity, religion, and gender laid the groundwork for a conflict that by 2004 had raged for more than twenty-five years.

Extremely poor and landlocked, Afghanistan saw its economic development pushed aside as a result of decades of war, nearly ten years of Soviet military occupation,[21] and the incomplete American invasion after 9/11. Economic disruption was severe in an agricultural economy that became one of the world's most land-mined territories.[22]

Labour shortages were chronic as population groups repeatedly fled the country. Pakistan and Iran at one time hosted more than six million Afghan refugees. Despite the repatriation of 4.2 million persons by 1 January 1999, the UN High Commissioner for Refugees (UNHCR) reported before 9/11 that the number of refugees fleeing to Pakistan due to war and drought had increased.[23] One million people moved internally to urban areas. After 9/11, the population again fled the cities toward the frontiers. War, population movements, loss of labour and investment capital, and disrupted trade and transport led to plummeting gross domestic product. Afghans have insufficient food, clothing, housing, and health care. In addition, the economy faces an enormous task in integrating persons with long-term disabilities incurred from land mines and warfare.

One measure of devastation is the humanitarian disaster for Afghan children. While human rights abuses hit all of society, the United Nations Children's Fund (UNICEF) says women and children were particularly harmed by:

(a) the fact that Afghanistan is one of the least developed countries in the world; (b) the long-term trend of high levels of mortality for both young children and women; (c) the war [that] has caused profound damage to the traditional coping mechanisms of families and communities; and (d) severe, institutionalized gender discrimination.[24]

UNICEF outlines especially troubling conditions for children. Infant mortality is 165 children per 1000 births. Maternal mortality is the second-highest in the world.[25] The mortality rate for children under five years old is 257 per 1000 births. Diseases that could be easily prevented by vaccines account for 21 per cent of child deaths, while diarrhea and acute respiratory infections kill another 42 per cent. Almost half of Afghan children are malnourished. Girls' education was banned in the 90 per cent of the country controlled by the Taliban, leaving only 36 per cent of boys and 11 per cent of girls in primary school. UNICEF says that drought left only 11 per cent of the rural population, or an estimated three million Afghans, with access to clean water. Drought also reduced cereal production in 2000 to 44 per cent of its 1999 level, forcing farmers and nomads to sell or slaughter herds and livestock. The move allowed survival but diminished rural self-sufficiency and compounded the negative impact of drought. International aid must address the humanitarian catastrophe before starting to promote economic development.[26] Self-sufficiency is all but impossible due to drought. Economic development is the ultimate solution to many of Afghanistan's problems but is deterred by lack of humanitarian assistance as well as continued insecurity.

One area of the economy that has flourished is the cultivation and production of illicit drugs. In 1999, Afghanistan was the world's largest opium producer and a major source of hashish. The CIA alleged that more and more heroin-processing laboratories were set up and all major Afghan political factions benefited from the drug trade.[27] A Six-Plus-Two Group[28] meeting in September 1999 expressed 'deep concern at the increased cultivation, production and trafficking of illicit drugs in and from Afghanistan and discussed establishment of a mechanism to enable them to cooperate more closely on counter-narcotics issues.'[29] The UN Office for Drug Control and Crime Prevention (ODCCP) formulated a regional plan to control the drug trade. Following a cultivation ban by the Taliban, the ODCCP reported in June 2001 that 'farmers in Afghanistan, the world's number one producer of opium poppy, did not plant the illegal crop this year.'[30] In the post-Taliban era, continued poverty makes opium a tempting source of revenue and production levels remain high.[31]

Successively a kingdom and a republic, the country's internationally recognized name is the Islamic State of Afghanistan (ISA). Overthrown by the Taliban in 1996, the ISA held the country's UN seat until an interim government took office on 22 December 2001. The Taliban Islamic Emirate of Afghanistan (IEA) was only recognized by Pakistan, Saudi Arabia, and the United Arab Emirates. The Taliban set up a Sunni Muslim fundamentalist regime that followed Islamic Law (Shari'a) by tacit agreement with all factions. After the ISA fled Kabul in 1996, the country had no functioning government structures. The Taliban focused on defeating other factions, consolidating power, and forcing international recognition as a de facto government until they were overthrown in the U.S. invasion of 2001.[32]

While the world focused on the Taliban's links to global terrorism, regional stability, refugees, human rights, the treatment of women, and the drug trade, less attention was given to the role of the international community in preparing the way for and sustaining the regime. Human Rights Watch (HRW) argued that

> the civil war in Afghanistan, a geopolitical battleground during the Cold War, is once again being sponsored by outside parties: Pakistan, Iran, Rus-

sia, and other neighboring countries with the United States and India working in other ways to influence the outcome.[33]

HRW called for a sustained arms embargo of all parties to the civil war, investigation of human rights abuses, an end to all military assistance, respect for international humanitarian law, and measures to defuse conflict so that reconstruction could begin and humanitarian issues be addressed.

The Taliban's global impact was confirmed on 9/11. Some Taliban factions supported Islamic militants worldwide and provided bin Laden with safe haven. Since Iraq displaced Al Qaeda on the U.S. policy agenda, human rights and humanitarian issues in Afghanistan have clearly suffered. U.S. policy statements recapitulate the heartland-thesis mindset in which the region is central for global security but its inhabitants are denied tools to build a stable and peaceful environment needed to achieve that end. After the Taliban declared themselves Afghanistan's legitimate government, their control of 90 to 95 per cent of the territory facilitated a revolutionary regime that resembled the Khmer Rouge and Sendero Luminoso in ruthlessness and hostility to the local population and the rest of the world. Until the question of legitimacy was resolved by the fall of the Taliban, the UN and Organization of the Islamic Conference left Afghan seats vacant. Ethnicization of the conflict was a key to the Taliban's defeat. Pashtun-dominated Taliban controlled Kabul and most of the country and were especially strong in the mainly Pashtun areas of the south while the opposition stronghold remained in the ethnically diverse north.[34] The Shia minority received Iranian assistance. Ethnic violence[35] was especially troubling since few institutions remain to channel conflict.

In addition to political organizations, several pressure groups are important. The hold of tribal elders, traditional Pashtun leaders, was seriously weakened by twenty-five years of conflict.[36] Afghan refugees in Pakistan, Australia, the United States, Canada, and elsewhere organized politically but did not influence the Taliban. Groups in Peshawar, Pakistan, included the Coordination Council for National Unity and Understanding in Afghanistan (CUNUA), the Writers' Union of Free Afghanistan (WUFA), and Mellat (a social democratic party). A diverse Afghan diaspora served as a contrast to the Taliban political system, which did not tolerate organized opposition or free expression of ideas. Opponents to the Taliban were imprisoned, tortured, and frequently executed.[37] Post-Taliban Afghanistan is characterized by a central government in Kabul supported by the UN and foreign troops, a redivision of the country through the rise of a system of regional and drug warlords, and continued violence.

The Afghan Mediascape

The Afghan mediascape under the Taliban included print media (magazines and newspapers) and electronic media (radio, telephones, television, and the Internet). Before the Taliban took power in 1996, telephone and telegraph service was very limited due to war damage. The Taliban repaired some international telephone links and re-established telecommunications between Afghan cities by means of satellite and microwave transmission in 1997. International telecommunications were provided by an Intelsat satellite earth station linked to Iran and an Intersputnik satellite earth station. A commercial satellite telephone centre was located southwest of Kabul in Ghazni.[38] Beyond telephones, the mediascape under the Taliban was very bleak. Television broadcasts, music, and images of the human body were outlawed in the name of Islamic 'purity.' In 2000, audiences for evening news on Taliban-controlled radio were as much as 70 per cent of the population. One radio station supplied the entire country with religious broadcasts (without music) and official propaganda. The ISA maintained national television broadcasting and regional stations in a few provinces. The Taliban converted the Kabul TV centre into a barracks.[39] The country had approximately 100,000 television sets before Taliban started a campaign to destroy them in 1998.[40] In 2000, the Taliban's information minister confirmed that the TV ban would remain in place.[41] By 2001, the only television broadcasts came from opposition-controlled Badakhshan TV, which aired evening news and endless loops of music, film, and cartoons.[42] TV rapidly returned after the Taliban fell in late 2001.

Freedom of expression did not exist in the Taliban mediascape. Newspapers were monitored and had to follow Taliban principles. No photos, features, readers' letters, or editorials were allowed. The Taliban allowed the English-language weekly *Kabul Times*, Pashto-language magazine *Nangarhar*, and Farsi-language newspapers *Hewad* (Fatherland), *Anees* (Companion), and *Shariat*. All news came from ministries or the official news agency. Journalists were under orders from members of the Taliban who were assigned to editorial offices. Paid badly and irregularly by the state, most journalists earned around 12 euros ($16.57 Cdn. in August 2001) per month. In July 2000 the government launched *The Islamic Emirate*, an English-language monthly to 'counteract the biased information put out by the enemies of Islam.' The first issue carried the front-page headline 'No terrorist camps in Afghanistan' and 'Extraditing Osama bin Laden would be scorning a pillar of our religion.'[43] Meanwhile, opposition newspapers in Pakistan and some Pakistani newspapers sympathetic to anti-fundamentalism printed Afghan opposition articles.

Foreign media were subject to Taliban control and image manipulation. In January 2000, CNN opened a Kabul office but had no right to film anywhere. Al-Jazeera also opened a Kabul office. The BBC and Reuters filmed Kabul street life so the world could see how UN sanctions impacted civilians, but operating under the Taliban was dangerous. In August 2000, Pakistani Khawar Mehdi, American Jason Florio, and Brazilian Pepe Escobar were arrested, questioned, accused of photographing a soccer match, and their film confiscated. After August 2000, foreign journalists received a Taliban 'good behaviour' list: no visits to private homes; no interviewing women without ministerial permission; and no photographing or filming people. Pakistani reporters in Afghanistan began to have trouble obtaining visas. BBC interpreter Saboor Salehzai was arrested in December 2000, charged with violating regulations for Afghans who work for foreign media, but released after four days.[44] The murders of journalists after the fall of the Taliban in October-November 2001 as well as the execution of the American journalist Daniel Pearl in 2002 illustrate the continuing tension in the region and the power of media in the conflict.

Aside from some administrators and VIPs, Afghans under the closed environment maintained by the Taliban could not access the Internet, which was a potential site for debate, ideas, and resistance. With ordinary Afghans unplugged by poverty, foreign intervention, civil conflict, and technophobia, the Taliban tried to use the Net to cultivate an international profile. However, an IEA website[45] set up to court global opinion was out of service by August 2001. The technophobic Taliban also had a website,[46] where a hacker posted porn, denounced Islamic fundamentalism, and left a Russian email address. A message at http://www.taleban.com with a photo of bin Laden said

Taliban was again fucked by RyDen. FuckZ: Usama Bin Laden, Ibn Al Khatab Maskadov and all chechen terrorists. BASTARD STILL WANTED DEAD OR ALIVE: 5,000,000$ for direct info. (it's not joke, its an official info from FBI). You can become a MILLIONAIRE, do it for yourself! Additional info about this you can find at www.fbi.gov. No more messages here, all people around the world already inform that are are only stupid monkeys Contact: ry_den@land.ru.[47]

Pakistan gave the Taliban authorities Internet access by hosting the Taliban and IEA websites. The Comsats server provided connections for several Taliban ministries, who apparently received 'preferential prices.' In 2000, a Taliban official in Peshawar told Reporters Sans Frontières (RSF) that he received email from the foreign affairs ministry and that several government branches used email. Aside from the Taliban, some NGOs and foreign correspondents accessed the Net via satellite. A photographer travelling in Afghanistan in 2000 said that assassinated opposition military commander Ahmed Shah Masood had an email address that he occasionally accessed via satellite phone.

Infrastructure in Afghanistan, which is still severely damaged, could hardly be more unlike Ireland's very modern facilities. The latter is a leader in the software industry and IT is linked to the dramatic transformation of that once-impoverished island. A striking feature of RAWA Web activism is that it concerns a country in which infrastructure and access are so severely lacking. Lack of infrastructure is a key feature of the specific RAWA context, which contrasts with IRSM Web activism in information-wealthy Ireland.

Post-colonial-feminism@activism.web

Based in Peshawar, Pakistan, the RAWA website embodies the Web's utility for resistance, alliance building, fundraising, and counter-discourse. Above all, the website serves as a powerful platform for *witnessing* events in Afghanistan under the Taliban and after its fall. The witnessing function is the website's most significant impact and a means by which Web activism shifts some attention from the conventional physical and coercive global-scape of conflict. RAWA thus illustrates both the rise of media tools in conflict and the increased prominence of gender issues in IR. Both trends are components of a wider transformation of politics and conflict. Left-of-centre RAWA radically challenged gender-based stigmatization and oppression by addressing global civil society on the Web. The website supplemented activism in physical space with a virtual context marked by representation, images, emotional-moral appeals, and a battle over perceptions. Just as RAWA, an independent political organization of Afghan women set up in Kabul in 1977 to fight for human rights and social justice, empowers the disempowered, its website facilitates activism that is impossible in territorial space. RAWA's electronic transgression of territorial politics started after civil war and the Taliban forced the organization to relocate to Peshawar.

The RAWA website focuses on the impact of fundamentalism on Afghan women. In August 2001, the homepage featured seventy-two reports, including appeals to the U.S. public through the *Oprah Winfrey Show* as well as to the UN. Press conferences, accounts of protests, denunciation of an alleged Afghan war criminal living in the UK, a statement on UN sanctions, photographs of atrocities, and information on RAWA activities were also posted on the site.[48] Alongside text, the site has extensive multimedia (audio and especially photo images) that enhance its moral-emotional appeal through images of Taliban brutality and a realistic portrayal of the oppression of Afghan women.[49]

The volume of hits on the site suggests that RAWA Web activism found a response in global civil society. On 22 August 2001, the website counter registered 824,617 hits. The number rose to 858,109 by 2 September 2001, an increase of 33,492. It rose 130,137 hits to 988,246 on 15 September (fours days after 9/11). On 16 September it reached 1,016,683 (an increase of 28,437 in twenty-four hours). By 19 September, hits increased again to 1,144,531, an additional 127,848. On 28 September, there was another increase of 415,281 hits to 1,559,812. The huge increase around 9/11 suggests that Web use follows practices on other electronic media,[50] given the latter's blanket coverage of events. Prior to this, the high volume of hits at the original homepage necessitated a mirror site. Unfortunately, there is no certainty as to who accessed the site, why, and how. The raw numbers of hits on the site climbed rapidly after 9/11 thrust Afghanistan and the Taliban to the forefront of global security debate and concern. Whether the increase was from journalists around the world verifying facts, individuals, government agencies, or some combination thereof was not clear. The point is that general interest in the site was manifest in the crude statistic of hits after 9/11. By 17 August 2003, the number of hits on the RAWA site reached 5,277,518.

Website Analysis – www.rawa.org

As in the case of the IRSM website, the contents of the RAWA website in August 2000 were used as a snapshot for this examination. The website's text, video, and audio resources are cross-referenced to draw visitors to specific materials. The homepage covers RAWA, its activities, the situation in Afghanistan, and suggestions for a global response. In general, documents are less than six pages long. The site is multilingual: English, Italian, Spanish, French, German, Farsi, Catalan, and Portuguese. In this analysis, English-language materials were employed to assess RAWA's use of a major international language and the potentials of global communication. The homepage is organized around eighteen categories: About RAWA; On Our Martyred Leader; Our Publications; Our Social Activities; Patriotic Songs (MP3); Photo Gallery; Reports from Afghanistan; On Afghan Women; RAWA Documents; RAWA Events; RAWA in Media; Movie Clips; Poems; Links; How to Contact Us?; How to Help Us?; Search Our Site; and Subscribe to Mailing List. The homepage is divided into theme areas and a features area. Effective use of colour, images, and sound communicate the brutal reality of Afghan women, illustrating the moral and emotional impact of immediate and authentic communication that uses globally transmitted representational imagery.

Website texts include reports, statements, press releases, and some longer documents. The features section contains media articles from sources such as ABC News, the *Toronto Star*, Tehelka.com (New Delhi), *Marie Claire* magazine, the BBC, and *La Vanguardia* magazine. Most reports on Taliban violence and human rights abuses centre on women and children. Abuses listed in the features section 'Recent Reports from Afghanistan' include public hangings; executions; arrests of foreign aid workers; destruction of TVs and VCRs; the Internet ban; population displacements; victimization of women in bride sales; destruction of rebel towns; school closings; religious intolerance against Shia Muslims, Sikhs, and Hindus; attacks on intellectuals; and widespread poverty. Many texts condemn the Taliban.

Another features section focuses on RAWA's political activities: a message of 'Afghan Solidarity'; press conferences; celebrations of International Women's Day; a demonstration at Kofi Annan's arrival in Islamabad; an appeal to suspend Afghanistan's UN seat; a statement about an Afghan war criminal living in the UK; photos of RAWA's French human rights prize; and a statement about UN sanctions against the Taliban. An additional section includes nineteen documents about RAWA's wide-ranging social activism: distribution of food, medicine, and blankets in refugee camps; photo documention of massacres; sale of RAWA t-shirts; participation in UNCHR sessions on human rights; home teaching in Afghanistan; campaigns in the United States, Canada, and Japan; and examination of the lives of those marginalized by the Taliban (especially beggars, prostitutes, and prisoners). Still another section has photos of atrocities, drought, refugee suffering, malnourished children, poverty, and fundamentalist terror.

'About RAWA' briefly outlines the organization's history and goals. Created to draw Afghan women into the struggle for human rights, social justice, women's rights, and democracy, RAWA says:

> the founders were a number of Afghan women intellectuals under the sagacious leadership of Meena, who in 1987 was assassinated in Quetta, Pakistan by Afghan agents of the then KGB in connivance with the fundamentalist band of Gulbuddin Hekmatyar.[51]

From early goals of a secular, democratic government in Afghanistan, RAWA broadened its interests to include education, health, and anti-poverty projects to secure gender rights and democracy.

After the Soviet-backed coup in April 1978 and the December 1979 invasion, RAWA joined the national resistance movement. However, RAWA's advocacy of democracy and secularism created distance between it and the Islamic fundamentalist 'freedom fighters.' Under Soviet occupation, RAWA activists worked in Pakistan with refugee women. They set up schools, hostels, a hospital, and women's nursing, literacy, and vocational training in Quetta, Pakistan. In 1981, RAWA began publishing the magazine *Payam-e-Zan* (Woman's Message) in Persian and Pashtu to raise social and political awareness among Afghan women. Some texts appear in English and Urdu. After the Soviet-backed regime was overthrown in 1992, RAWA's struggle turned to fundamentalism, Taliban atrocities against all Afghan people, and the latter's ultra-chauvinism and misogyny. RAWA has a long-standing goal of broadening its social and relief work, but 'unfortunately we do not at the moment enjoy any support from international NGOs, therefore our social programmes are presently greatly reduced for lack of funds.'[52]

In 2000, the 'About RAWA' category included a May 1995 Amnesty International (AI) document, 'Women in Afghanistan: A Human Rights Catastrophe,'[53] and a March 1999 AI call to protect an International Women's Day demonstration by Afghan women in Pakistan. A third AI text in April 1997 condemned the torture of a RAWA sympathizer in Pakistan. Another link documented the presentation of The

French Republic's Liberty, Equality, Fraternity Human Rights Prize to RAWA on 15 April 2000. A final link in the category specifies RAWA's aims:

- struggle against the Taliban and Jehadi types of the fundamentalists and their foreign masters.
- establish freedom, democracy, peace and women's rights in Afghanistan.
- establish an elected secularist government based on democratic values.
- unite all freedom-loving and democratic forces and to struggle against all those who collaborate with the fundamentalists.
- struggle against those traitors who want to disintegrate Afghanistan by causing tribal and religious wars.
- launch educational, health care and income generation projects in and outside the country.
- support the freedom-loving movements all over the world.[54]

'About RAWA' is posted in Persian, French, Italian, Spanish, and German.

'On Our Martyred Leader' features a biography of Meena, RAWA's first leader. Born in 1957, Meena became an activist in the 1970s. After the Soviet invasion, she organized meetings and protests in schools, colleges, and at Kabul University. RAWA says her main accomplishments were starting *Payam-e-Zan* and Watan Schools for refugee children. Meena's biography is posted in Persian, Italian, Spanish, and French. The section also contains MP3/WAVE recordings of Meena, a translation of one of her poems, photographs, a song in English that is dedicated to her, and an article about her life and struggle from the March-April 1999 American girls' magazine *New Moon*.

'Our Publications' features links to past and current articles in *Payam-e-Zan* in Farsi, Pashtu, and Urdu.[55] It also contains the English-language publication 'The Burst of the "Islamic Government" Bubble in Afghanistan,' consisting of 122-page and 96-page sections.[56] In addition a twenty-four-page colour brochure outlines RAWA activities, viewpoints, and the Afghan situation. Sixteen other items posted in the category include a seventy-two-page text, 'Afghan Women Challenge the Fundamentalists,' four posters, a sticker, special bulletins for International Women's Day and demonstrations, greeting cards, audiocassettes of patriotic songs, AI reports on Afghanistan in Pashtu and

Farsi,[57] and t-shirts. The items are sold to finance RAWA assistance to Afghan refugees and show how the organization also understood that the Web is a marketing tool. One analytical issue is whether Web activism will become a political marketing tool, a platform for activism, or some combination of the two. Marketing could be seen as a component of activism if the latter is broadly understood as a process of trying to convince a public of the value or need to enact a platform or spread a set of ideas.

'RAWA's Social Activities' outlines events in Pakistan and Afghanistan as well as future plans. RAWA activities in Pakistan include primary and secondary schools for children, literacy courses for women, an orphanage, health care in refugee camps, and a hospital that is near closure for financial reasons. In addition, RAWA provides news and reports on Taliban and fundamentalist abuses to human rights NGOs and media, and produces anti-fundamentalist and educational cassettes, plays, poetry, and propaganda. It also organizes demonstrations and press conferences, issues press releases and leaflets, maintains the website, and sets up media contacts and interviews.

In Afghanistan, RAWA supports female victims of war and atrocities. Families who suffer violence are contacted so that their accounts can be published in *Payam-e-Zan* or AI can be alerted. If post-traumatic counselling is feasible, the traumatized family or children are moved to Pakistan, missing family are located, basic needs are supplied, or sponsors are found. Regular RAWA activities in Taliban Afghanistan included education-propaganda (home schools and literacy courses, discussion groups on women's rights, education, democracy, and civil liberties), health care (health teams in seven provinces treated women who could not visit doctors for fear of the Taliban or due to poverty; treatment for children and the wounded was provided, as well as first-aid classes and polio vaccinations), and economic aid (chicken farms, carpet-weaving, embroidery, knitting, bee-harvesting, handicrafts, tailoring, short-terms loans to widows and families). RAWA wants to expand its activities to include more education (computer, English, and trades courses for women and girls), publications in main Afghan languages, and magazines that address 'taboo' subjects.

'Patriotic Songs' and 'Photo Gallery' employ the Web's multimedia capacities by means of MP3 music and photographs. 'Patriotic Songs' includes twenty-one songs in Pashtu and Dari, and one in Urdu. The music ranges from songs for Meena, condemnations of Islamic fundamentalism, appeals to fight for democracy and freedom, laments for

Kabul, to children singing in a RAWA school. In 2000, 'Photo Gallery' linked fifty-two photographs of Afghan society, the impact of conflict, and RAWA activities. Photos document Taliban massacres, hangings, and mutilations as well as poverty and drought. Photos of RAWA activities cover rallies, home-based classes, and the distribution of food and blankets to refugees.

'Reports from Afghanistan' features nine types of documents, including reports from Afghanistan in English and Persian. Also available are documents on what RAWA calls 'fundamentalist criminality,' the condition of women, fundamentalist destruction of Afghanistan's heritage,[58] two *Payam-e-Zan* reports, a report from the city of Herat about life under Taliban, and a photo gallery. At this point, the website is complicated by cross-links between different categories. Links to 'Afghan Women under the Tyranny of the Fundamentalists' and 'Photo Gallery' are also categories on the homepage. The difficulty in managing large amounts of information on small budgets and limited personnel is evident. Clearly, eased global communication raises new complexities in terms of information overload, management of documents, and comprehension of huge volumes of material. The danger is that key information could be 'buried' in a mass of texts. RAWA obviously weighed the merits of posting or not posting information, and decided to provide it in spite of this risk.

'On Afghan Women' opens to 'Afghan Women under the Tyranny of the Fundamentalists,' with texts and photos of Taliban atrocities. At the top of the page, RAWA provides an overview of women's conditions and restrictions.[59] Information comes from various sources. 'Inside Afghanistan: Behind the Veil' presents journalist Saira Shah's account of a visit to the country to conduct interviews and take photographs for a June 2001 BBC News report.[60] Other documents describe the conditions for Afghan women in a society in which they could not work or be personally secure away from home. Some conditions listed include rising drug addiction; a ban on female university students, civil servants, and teachers; trafficking in women; murders; rapes; house arrests; floggings; public executions; restrictions on female foreign aid workers; polygamy and concubinage; beatings; and oppressive clothing. Photographs depict a mother of four who was raped and killed by fundamentalists, an Afghan widow, a woman forced into prostitution by poverty, women forbidden to attend a funeral, a woman speaking about her husband's death, public execution of a woman, and the wounds of an elderly woman after attack by fundamentalists. The images document Taliban abuse and highlight gender's relevance to IR as a specific form of human rights violation.

'RAWA Documents' incorporates forty-one texts. These include statements on the fifth anniversary of the Taliban's conquest of Afghanistan, condemnations of the Taliban, UN sanctions and Pakistani border closures, commemorations of International Women's Day and International Human Rights Day, and a message to the Association of University Women in California. 'RAWA Events' archives group activities after 1998. Events listed include a press conference after cancellation of a 'Black Day' (28 April) demonstration,[61] a RAWA rally attacked by fundamentalists, RAWA rallies in Islamabad and Washington, DC, and sit-ins and demonstrations. Documents include text, photos, movie clips, articles reproduced from the press, and Persian translations. 'RAWA in Media' lists dozens of documents from various media, including English- and Spanish-Catalan–language sections. The English-language section has articles from media in the United States, UK, Pakistan, Japan, India, Brazil, South Africa, Portugal, and France, as well as from international news agencies like AP and Reuters. An article from *Le Monde diplomatique* is reproduced in French while others are in Portuguese. The Spanish-Catalan list has articles from Spanish media.

'Movie Clips' provides various short videos[62] in MPG, AVI, RealVideo, RM, and GIF formats in three galleries. The first focuses on brutality and violence in Afghanistan. Highly dramatic clips include the execution of a woman in a stadium with a rifle to the head, the slitting of prisoners' throats, mass graves, and the use of building cranes for public hangings. The second gallery shows social devastation: ruins in Kabul, women begging, and people forced from homes. The third gallery has images of destruction due to fighting between Taliban and Jehadi fundamentalists: a mosque, Kabul, and houses. The third gallery also has clips of RAWA home classes inside Afghanistan and a large RAWA demonstration in Pakistan. Below the three galleries, clips document RAWA events and rallies. 'Poems' includes those written about exile, Meena, deaths in the conflict, women's rights, and freedom and democracy. The page also has hyperlinks to RAWA song lyrics.

The 'Links' category contains many hyperlinks, although several were not active by 2001. Hyperlinks appear to have been selected due to their international profile (e.g., the UN and AI), country of origin (eleven from the United States, two from international organizations, and one from Australia), and general perspective on Afghanistan (i.e., anti-fundamentalist and pro-RAWA). A 'Jubillenium' banner at the top of the page[63] is inactive. 'Amnesty International Reports on Afghanistan'[64] leads to a page containing all AI materials since 1996. 'United Nations Report on Human Rights Situation in Afghanistan'[65] leads to a 20 January 1995 report by the UN Commission on Human Rights Special Rapporteur Felix Ermacora, 'Final Report on the Situation of Human Rights in Afghanistan.' The next link opens on 'FemAid.org,'[66] an organization geared to practical responses for women's rights which helped women in Bosnia. It has a special page on Afghanistan. One page[67] calls for sponsorship of teachers in Afghanistan as part of a RAWA-Afghan Women's Mission project. It also provides information about Azadi Afghan Radio[68] and Calgary-based Women for Women Canada.[69] The Afghan Women's Mission asks for donations to rebuild RAWA's Malalai Hospital for refugees in Pakistan, a project that is supported by the Simon Weisenthal Center.

'Acting in Solidarity with Afghan People (ASAP)'[70] is a California-based aid organization that posts reports and news articles on Afganistan and human rights abuses. 'Octaves Beyond Silence' is another California-based non-profit group that uses music and art to help female victims of violence.[71] In December 2000, it released a compilation CD, the proceeds from which are used to assist women. 'The Afghan Women's Mission (AWM)'[72] leads to yet another California-based organization that aids women and children by raising funds to reopen Malalai Hospital. The California-based 'Dr. Homa Darabi's Foundation, USA' aims to raise awareness of the 'violence afflicted on women and children all over the world in the name of God, religion, culture, family values, and the preservation of society, and to promote activism for change.'[73] The variety in this section shows how Web activists use hyperlinks to indicate their political stances and create communities of interest and/or support around particular issues by highlighting like-minded groups. The links also use more conventional political techniques, such as petitions and appeals for funds. The Web makes it possible for even socially marginalized groups in remote locations with poor infrastructure to engage in some of the same activities as mainstream political organizations in North America and Western Europe. In this case, global awareness of RAWA's issues accompanies a worldwide appeal for assistance for the organization.

The 'Afghan Community in Australia'[74] site connects Afghans of all ideologies, religions, and ethnicities in Australia, New Zealand, and the Asia-Pacific region. It provides news; religious, cultural, and historical information; news about Afghans in Australia; as well as Afghan music, cuisine, and photos. 'News on Afghanistan from Reuters' leads to varied news headlines (from ESPN.com, ABCNEWS.com, Disney.com) with no direct bearing on Afghanistan. 'News on Afghanistan from AFP' and 'News on Afghanistan from Yahoo News' are dead links. 'News on Afghanistan from Excite News' activated a search that produced eighty-six results. 'Daily News from Afghanistan' was not found while 'Afghanistan News Archive' was linked to a search engine.[75]

The site 'National Organization for Women, USA (NOW)' features a link to global issues.[76] 'Stop Gender Apartheid in Afghanistan!'[77] leads to a page maintained by the Feminist Majority Foundation to pressure the U.S. government and UN to help restore Afghan women's and girls' rights. It includes news, an online petition, and calls for donations. 'Golshan Society (Persian)'[78] leads to a bilingual Persian-English site that describes itself as progressive. The English page is entitled 'Islam Unveiled with Rational Thinking' and includes a puzzling statement by 'Ali Sina' denouncing Islam as a dangerous, intolerant fraud of 'sincere' Muslims. An attempt to link to http://www.freespeech. org/CoxNews/ led to '403 Forbidden.' At the bottom of the links page, major search engines help locate more Afghan information.

'How to Contact Us?' lists the ways to get in touch with RAWA. Addresses for a post office box in Quetta, Pakistan, and AWM in Pasadena, California, are provided. Below this, another page explains how to donate funds to RAWA: cheque or money order to AWM in Pasadena, California; cheque or money orders in U.S. dollars to Mrs Sohaila Farhad in Quetta, Pakistan; credit card or PayPal; or electronic funds transfer directly to RAWA in either U.S. dollars or deutsche marks.[79] RAWA may be contacted by email at rawa@rawa.org, by telephone in Pakistan, or fax in the United States or UK. RAWA manages a mailing list that allows visitors to send messages as well as snail mail and electronic addresses.

'How to Help Us?'[80] provides various ways for the public around the globe to assist RAWA. In addition to donations, visitors may sign an electronic petition.[81] Fundraising and awareness drives to assist RAWA have been organized on Quadra Island, British Columbia, in Santa Barbara, California, and by northern California high school students.[82] Other ways to take action are: demand that the British government investigate and prosecute a war criminal living in the UK and send donations to impoverished refugees in Peshawar, Pakistan.

Another page requests donations to reopen Malalai Hospital, which specializes in land mine injuries.

RAWA lists many ways in which the public can help: introducing it to individuals, groups, schools, organizations, and congregations; staging protests and demonstrations to support it and Afghan women; organizing meetings and seminars to highlight the situation in Afghanistan; writing to Pakistani authorities to protest government and non-government violence against RAWA; inviting RAWA members to speak; covering Afghanistan and fundamentalist crimes in publications; translating RAWA materials into English and website materials into Spanish, French, Portuguese, Italian, and Arabic; distributing RAWA publications and audiocassettes; sending money and supplies for RAWA schools; organizing fundraising campaigns; sending medical supplies; donating computers, camcorders, cassette duplicators, sound and film mixing equipment, CD recorders, small photo and video cameras for RAWA documentation of fundamentalist crimes, and miniature video and photo cameras to film and photograph atrocities in Afghanistan; distributing films, books, and materials with progressive and anti-fundamentalist themes; and sponsoring teachers in Afghanistan.

RAWA's use of the Web for fundraising and lobbying demonstrates that IT has a 'ripple' effect, as its impact extends beyond a direct territorial focus. In this case, an Afghan feminist organization in a Pakistani refugee camp was able to raise funds globally. RAWA's success is that it went beyond the very limited possibilities of Afghanistan to appeal to global civil society. RAWA in this respect shows how Web activism blurs conventional measures of political effectiveness such as fundraising and petitions. The conventional view is that, in the end, only political behaviour that resembles that in Western democratic systems is efficacious and the stuff for analysis by IR, political science, and other social sciences. The problem is that a non-state actor such as RAWA exists in a postcolonial society in which these activities are not an option. As such, RAWA fundraising and petitions suggest how IT is influencing global politics. This means that an evaluation of RAWA in a conventional sense is not particularly useful except insofar as it touches on its Web activism.

As the discussion above regarding hits on the RAWA site illustrates, the issue of data is an important one in analyses of IT. The problem is that there are very few reliable sources of public data on access to specific political Web sites. For that reason, it is difficult to relate existing data to political outcomes. This analysis is not geared toward providing data on Web use. Instead, it shows how specific non-state actors use the Web both to respond to their context and to appeal for support on a broader basis. By examining a selection of non-state actors from Northern Ireland, Afghanistan, and Peru, this analysis suggests both the diversity of organizations that use the Web for political purposes and the barriers to Web activism. The limits on Web use due to gender, race, and ethnicity to a great extent motivated my interest in how once-marginal non-state actors are using the Web.

Web Polity, Web Combat?

In a mediascape characterized by far-reaching practices of globalization and IT innovation, RAWA projects issues of gender-based oppression from their territorial locale into global conflict. While RAWA's main concerns are gender-based discrimination and oppression, anti-female violence, brutality, and the killing of women and children by the Taliban, the organization also concentrates on the deep-rooted poverty, fundamentalism, inequality, and violence in Afghan society. RAWA's solution is a democratic, secular regime committed to social justice. It sees both the Taliban and the Northern Alliance that replaced it with U.S. support in 2001 as the results of foreign intervention whose recourse to religious law is especially oppressive for women and a factor that sustains violence, socio-political disintegration, and ethnic conflict.

Describing itself as committed to freedom, democracy, secularism, and women's rights, RAWA advocates unity against groups and individuals allied to fundamentalists. It does not call for violence, but aims to generate awareness and spread its views through education, membership drives, and Web activism. Characterizing Afghan women as politically and socially oppressed, RAWA opposes foreign intervention in Afghanistan and supports liberation movements in Palestine, Kurdistan, Kashmir, Iran, and elsewhere. RAWA sees itself as resembling these movements because it is also struggling to free Afghanistan from the influence of foreign Islamic fundamentalists in a conflict that resembles struggles for national self-determination.

The RAWA website raised awareness of its struggle by communicating information on political and social conditions under the Taliban and ISA to friends and supporters. With limited opportunities to engage in domestic activism under the Taliban or the ISA, RAWA used

Web activism to circumvent oppression. Given the often hyperbolic claims as to the impact of IT, the issue is the extent to which Web activism advanced the cause of Afghan women and produced results. Various perspectives exist as to the impact of IT. For example, the hyperbolic cyberutopianism of Nicholas Negroponte's *Being Digital*, widely read and cited when it appeared in the 1990s, argued that the Net would foster a revolutionary transformation in the way that politics and society are conducted by placing powerful communicative devices in the hands of individuals and groups.[83] At the other end of the spectrum analysts such as Paul Wilkinson[84] or Bruce Hoffman[85] argue that IT places powerful abilities for communication in the hands of dangerous terrorists and other radicals. In the event, RAWA is a non-state actor with limited means and a non-violent agenda. Its use of Web activism produced modest results if measured in terms of geopolitics or state-centred public policy. RAWA repeatedly emphasizes that finances hamper effective lobbying and participation in demonstrations. In response, Web activism has promoted global awareness of the oppression of Afghan women and helped build ties to other women's groups and individuals. For this reason, RAWA's greatest contribution was to *witness* events in Afghanistan and to spread that information to global civil society.

RAWA's insurgency online is based on a distinct form of Web activism. As will be seen in the next chapter, it is unlike the MRTA website insofar as it has fewer links, emphasizes fundraising, and duplicates some materials. RAWA's Web activism is based out-of-territory in an adjacent country where activists' status is tenuous, while the MRTA English-language site was run by supporters on another continent. RAWA's Web activism employs more multimedia, photos (uses shock value of human rights abuses), and highly dramatic video clips. This use of multimedia has facilitated global witnessing. The MRTA and RAWA are also different types of groups. Unlike the MRTA, RAWA is a non-violent organization active in an extremely violent context and has itself been the target of fundamentalist violence. Given RAWA's distinct use of the Web, both groups illustrate the variety of Web activism in the contemporary mediascape as well as the need for further research.

RAWA's Web activism also differs from that of the IRSM. The IRSM site is more text-based, is oriented to specific political debates, and uses a nationalist ideology that is much more profoundly rooted in national history than is Afghan feminism in its own context. Critically,

IRSM Web activism aims at a specific group of organizational members and ideological adherents that it aims to consolidate into a network. RAWA's public is more general and global. The IRSM is aligned with foreign groups in like-minded pursuit of territorially based political ends, while RAWA's main external supports have been Californian and Canadian feminists concerned about the global impact of gender oppression.

Global witnessing also gives RAWA's website a more archival-documentary character than the strongly ideological sites of the MRTA or ISRM, which are more directly linked to a quest for political power in physical space. Global witnessing documents particular circumstances that relate to a group's motives. It is key for understanding the impact of RAWA because it provided both a raison d'être for and their contribution to postcolonial feminism. Historically, gender issues have been poorly documented since women have been at a disadvantage in terms of assembling historical documents. Weimar German feminism was so completely eradicated by the Nazis that feminism did not appear again in the former Federal Republic until the 1970s. Historical continuity was completely broken. RAWA effectively circumvented this type of catastrophic possibility by placing its documents online. Web activism makes it difficult to oppose or dismiss RAWA by arguing that gender politics do not exist in Afghanistan or are a foreign import to a traditional society.

Unlike the IRSM website, RAWA's insurgency online extensively and consciously uses the hypermediating potential of new global IT. This is especially apparent in the incorporation of video and photo materials. RAWA uses various textual and non-textual media to send a message to global civil society, including poems, newspaper articles, coffee cups, t-shirts, and posters. Despite severely limited physical circumstances under the Taliban, RAWA used representation and imagery to connect to other media. In this way, the organization transformed itself into a media product, a form of tech sophistication that strikingly counterpoises the devastation, sectarian violence, and technophobia in Afghanistan. The MRTA website also aimed at mainstream global media, but its highly ideological content differed from RAWA's documentary character.

While the Taliban remained in power, RAWA's Web activism was an important site for global witnessing, a venue for criticizing and resisting a totalitarian ideology. In light of the success of twentieth-century totalitarian regimes in crushing indigenous feminism, the fact of witnessing is a significant achievement. The huge number of hits on the site suggest that its influence in global society was real. Global society could ignore RAWA but not the plight of Afghan women. RAWA's website is too complete and powerful. Its audio and visual media created an interactive springboard for a moral-emotional appeal, enhanced by the use of numerous testimonies. The personal experiences of women and other witnesses and extensive cross-documentation of accounts provide an account of a social and political tragedy that resonated throughout the world.

Given that RAWA has had to operate in Afghanistan with exceptional caution, that its target was the Afghan diaspora and global civil society, and that the Taliban also tried to maintain an extra-territorial Web presence, the peculiarities of the case are evident. RAWA operated its website on the basis of a global vision, especially aimed at the United States, the UN, NGOs, and sympathetic activists and donors. The nature of the Afghan conflict denied the physical props of traditional politics to RAWA. In this sense, the RAWA website uniquely demonstrates the relevance and applicability of new IT in a postcolonial setting, in a society that has experienced profound trauma, dramatic disenfranchisement of major segments of its population, and cynical manipulation by the great powers. The website thus encapsulates the limits and promise of electronic representation. The limits imposed on RAWA are many: inability to conduct open politics in a physical space; no direct access to victims; lack of active above-ground organization in Afghanistan; extreme danger; deficient funds; and illegality in its territory of origin. On the other hand, RAWA embodies the promise of electronic politics: the articulation of a counter-discourse to a savage regime, an appeal to global civil society, the ability to witness and to resist by 'thinking otherwise,' and cross-organizational, cross-ideological appeal.

In light of the Taliban's savage misogyny, the most remarkable feature of the RAWA website is its existence and relevance to a conflict with global import. It emerged in a society in which women began to break away from traditional social structures in the last forty years of the twentieth century. The society experienced a coup d'état, revolution, foreign invasion, civil war, counter-revolution, and foreign intervention in a cycle from 1973 to 2001, a short time frame of twenty-eight years. The position of Afghan women is a key to the conflict, as women entered professions, assumed leadership roles, and then saw advances brutally rolled back. As Afghan women's voices were privatized by the

Taliban, the existence of a very public medium was a major achievement. RAWA Web activism alerted global society and motivated the Taliban's efforts at an electronic response. When http://www.taleban.com and the site of the Islamic Emirate at http://www.afghanie.com were unavailable in August 2001, it was clear that RAWA had won the electronic civil war.

Ironically, a sign of the power and relevance of insurgency online is that the public about which the RAWA website speaks (Afghans in Afghanistan in general and women in Afghanistan in particular) was not even online! RAWA shows the relevance of the Web as a tool for disenfranchised, silent, or exiled political movements. In this case, RAWA found a tool to fight a neo-totalitarian regime, bypass location, alter the nature of security, as well as rattle the cage of postcolonialism - that is, to overcome the limits of a situation in a marginal society that has been more acted upon than active in global politics.

4 Insurgency Online as Media Relay: The Web Activism of the MRTA[1]

As the RAWA and IRSM websites illustrate, Web activism is carried out to many ends by a range of non-state actors in both democratic and non-democratic societies. A common characteristic of the three online insurgent organizations discussed here is their call for far-reaching change in their host society and radical opposition to the existing political regime. They carry out politics in a transnational manner that highlights the relevance and reach of non-state actors in a post-realist global mediascape. In democratic societies, non-state actors as diverse as white supremacists, animal rights activists, or ethnic nationalists also use Web activist techniques to reach a global public. In some non-democratic settings, insurgency online sometimes resembles classic North American–West European liberal democratic ideologies, but religious, gay and lesbian, Marxist, and other movements are also very active. In 2002, for example, al-Qaeda used the Web to spread 'videos of terrorist attacks, proclamations by al-Qaeda's leaders and call to Muslims to take action against the West.'[2] Non-state Web activists share an ability to send political messages to a wider local, national, or global group regardless of government policies, politics, or political culture in their physical location. Alongside the pioneering Web activism of Mexico's Zapatistas, the Peruvian MRTA was an early movement to embrace Web communication in its fight for social justice against an authoritarian government.

MRTA Web activism differs from that of the IRSM or RAWA because it accompanied a violent guerrilla war with the Peruvian government. In spite of MRTA's physical force tactics, information terrorism does not accurately describe insurgent movements that are active *information providers* using Web activist techniques. Web activism describes the wide range of non-state actors who understand how global power is linked to the ability to articulate, organize, and communicate information over distance.[3] The case of the MRTA shows how violent non-state actors also exercise the power of information. In interpreting its significance, MRTA 'Web activism' is conceptually distinct from information terrorism and redirects analysis to the implications of information provision for global communications. A significant implication of Web activism is the connection between the Web and a new 'public sphere'[4] in which ideas, debate, and public intervention[5] occur in a non-physical 'place.' The public sphere is formed through the independent representation of non-state actors by images and text that autonomously communicate values, interests, and needs. Rather than simple propaganda, Web activism embodies the altered nature of the post-1989 global-scape in which broader social, cultural, and political factors converge in ways that cannot be sufficiently conceived with the categories of bipolar and state-centric political imagination.

As RAWA and the IRSM suggest, insurgency online is shaped by the interconnected informational and communicative aspects of the Web. The information and communication character of the Web in turn conditions how groups globally transmit political messages. Communication presupposes community, which makes the Web a focus for community development even though the resulting ties are partly non-physical or 'virtual.' Community and communication are intimately linked 'because it is only through communication that values, for example, can be shared and made common to the group.'[6] Given this connection, MRTA Web activism embodies and articulates a sense of community alongside a dominant physical-place political culture. MRTA Web activists used the Web in order to construct of series of networks with media, militants in other countries (the United States, Canada, Italy, France, Denmark, Germany, etc.), and their supporters. The network took on its own dynamic. Although MRTA militants who seized the Japanese embassy in Lima in 1996–7 were killed, a parallel Web community successfully continued its campaign to discredit the Peruvian government. Information – that is, the *content* of communication – spread specific MRTA Web activist messages. The organization's activities underline how information '*itself* is conditioned and structured by the social institutions and relations in which it is embedded.'[7] MRTA Web activism is insurgent and political, and expresses Peru's long-standing social fragmentation. Unlike more widely known forms of contemporary global information (such as news or economic data),

MRTA Web activism was not created as a commodity. Yet, through the embassy crisis, the global news industry transformed information on the website into a consumable thing. At a more fundamental level, the power of communication and information in Web activism works as a narrative that expresses value(s) and provides the raw materials to construct political identities. In the case of the MRTA, Web activism was part of a broader social movement that called for the removal of the authoritarian Fujimori regime.

Since MRTA Web activism explains, justifies, and rationalizes political violence, it raises a series of conceptual, policy, and civil rights issues. The group's narrative, like those of many other non-state actors, aimed to discredit and pressure the Peruvian government.[8] The significance of the MRTA website is not 'technical training, but training in using the technology in a strategic way.'[9] In Peru, MRTA Web activism was a component of globalization that had a transnational character and impact. It paradoxically served as a catalyst for intensified ideological struggle (the battle for 'hearts and minds') while partially deterritorializing national issues. As part of a mediascape that is both global and local, Web activism feeds older media practices such as radio, television, and newspapers. Where the mediascape does not provide groups like the MRTA with an independent capacity for interaction, the Web opens possibilities for informing and influencing a public by articulating a point of view and developing autonomous representations of events. As with other global media (e.g., satellite television or cellular telephones), the impact of the 'MRTA Solidarity Page' is linked to its role as a platform for messages that transcend domestic political boundaries.[10]

The Peruvian Mediascape – Conflict, Fragmentation, Corruption

Left-wing guerillas and successive Peruvian governments waged a civil war in the 1980s and 1990s. In 2002, a Truth and Reconciliation Commission (TRC) was set up to investigate atrocities that were committed by both sides. In its August 2003 report the Commission estimated that 69,280 people had been killed in the conflict. The civil war erupted following the launch of a guerrilla struggle against the Peruvian state by the extreme-left organization *Sendero Luminoso* (SL – Shining Path). During the conflict, violence fell unequally on different regions and sectors of Peruvian society, recapitulating its historic inequalities and divisions. The largest number of victims were among the peasantry, the poor, and the uneducated, which shows 'in the TRC's judgment, the veiled racism and scornful attitudes that persist in Peruvian society almost two centuries after its birth as a Republic.'[11] Being poor, rural, and illiterate, in other words, not only marginalized segments of Peruvian society, but also put individuals at risk from the media-savvy segment of society that seeks, by various means, to 'develop' the disadvantaged. More than anything, the conflict demonstrates the failure of state formation and development in Peru. The results of the process are structures that can neither provide security nor guarantee the basic rights of citizens.

Peru is a deeply fragmented society that has lacked effective government and social cohesion for most of the period since the Spanish explorer Francisco Pizarro conquered and pillaged the Inca Empire in 1532–3. The split between European and indigenous societies in Peru dates from this period. Other fractures, such as that between the coastal plains, mountainous interior, and Amazon jungles, were exacerbated as the economy developed dependencies on globally traded commodities that include guano, minerals, and illicit drugs, depending on the time period. The emergence of an industrial economy along the coast contrasts with the deeply seated poverty of the indigenous agricultural sector. Neither has provided a firm basis upon which to establish either a coherent national identity or effective governance. Peru is a country that has experienced globalization as a catastrophe, whether in the form of *conquistadores* or International Monetary Fund (IMF)–led economic restructuring. In the 1980s and 1990s, both SL and the Tupac Amaru (MRTA) led separate and antagonistic guerrilla wars against the state that themselves further illustrate Peru's fragmentation, in this case among opposition movements.

By the late 1990s Peru was reeling under the impact of civil war and the authoritarian rule of Alberto Fujimori (1990–2000). In this context, the role and operation of media as a fulcrum for debate, information, and the expression of political, cultural, and social identities, were especially precarious. In contrast to Northern Ireland, where the media-scape was well developed, and Afghanistan, where the information economy was destroyed by twenty-five years of conflict, Peru has many print and electronic media that themselves became directly involved in the civil war. Fujimori and his notorious henchman Vladimiro Montesinos systematically used corruption to control media, especially television. One of their principal means of control was 'vladivideos,' secret recordings made by the former head of the secret police (Servicio de Inteligencia Nacional – SIN), Montesinos, when he handed money to politicians or businessmen. Some videos show large amounts of money going to heads of TV stations to ensure support for Fujimori's 2000 election campaign. Montesinos also gave $3 million to Samuel and Mendel Winter in 1999 so that they could win control of Frecuencia Latina TV from Barach Ivcher, an Israeli-born Peruvian citizen who had seriously criticized the Fujimori administration. It was only in 2001 that the Supreme Court acquitted Enrique Zileri Gibson, signalling a return to normalcy in the Lima press. Gibson, managing editor of the weekly *Caretas*, was sued for libel by Montesinos in 1990 after he ran an article referring to the latter as 'Rasputin' because of his closeness to Fujimori.[12]

Throughout his presidency and at the time the MRTA seized the Japanese embassy in Lima, bribes, secret recordings, and libel charges were only some of the tools used by Fujimori to ensure media compliance. In 1997, for example, army intelligence agents tortured one of their own subordinates, Leonor La Rose Bustammante, whom they suspected of leaking information to the media. In a dramatic hospital bedside interview on *Contrapunto*, a Channel 2 television program, La Rosa showed severe burns and scarred fingernails, and could only walk with assistance. The same program also investigated the 'disappearance' of Mariella Lucy Barreto Riofano, another SIN agent and friend of La Rose. Her dismembered body had been found in plastic bags on a road near Lima. On 1 April 1997, armed men seized Blanca Rosales, general editor of the left-of-centre opposition newspaper *La République*. She was beaten, but managed to escape. On 1 June 1997, César Hildebrandt, presenter on the television program *En Persona* that aired revelations by La Rosa, received a telephone death threat against his son. Soon after, three armed men in military-style clothing attacked an *En Persona* film crew. Still later, three armed men beat up the political editor of the newspaper *Ojo*, Luis Angeles Laynes. Although the government said that it had no responsibility for such attacks, 'it openly denounced journalists and media proprietors who published unwelcome disclosures, and opened selective prosecutions against them for alleged tax debts.'[13]

As the TRC report notes, media also sometimes functioned in a manner that perpetuated rather than diffused civil conflict:

The TRC has found that in many instances news media fell into crude presentation that was inconsiderate to the victims and offered little to inspire national reflection and sensitivity to the issues. Part of this problem was the implicit racism of the media ... the issue of subversive and counter-subversive violence was not treated in a way that would entail a significant contribution to the pacification of the country.[14]

When the MRTA seized the Japanese embassy on 17 December 1996, both foreign and national media focused on the crisis and became caught up in events. Fujimori was furious when fifteen journalists entered the compound. A journalist from Japan's Kyodo News Agency, Koji Harada, brought some of the first images of the guerrillas to the world. The government had tried to cut media off from the guerrillas by shutting down electricity and telephone lines to the compound. On 7 January 1997, Peruvian police detained a Japanese reporter from Asahi TV, New York–based Tsuyoshi Hitomi, and his interpreter, Victor Borja, after they surreptitiously crossed police lines for a two-hour interview with the MRTA. Their notebooks and camera were confiscated. The correspondent for London's Worldwide Television News (WTN), Miguel Real, had to leave Peru due to death threats after his short wave radio interview with MRTA leader Nestor Cerpa. The government's leading spokesperson in the crisis, Domingo Palermo, complained that foreign reporters were giving advice to the MRTA leaders.[15]

The MRTA

In Peru's fragmented and conflict-ridden environment, the MRTA rejected parliamentary politics and situated itself in the indigenous tradition of anti-colonial liberation struggle. The group went underground and structured itself into armed units shortly after it was created by radical-left organizations in 1984. In contrast to SL's imposition of 'mass struggle' on the population, the MRTA called itself an organization of 'the people' that is building a social coalition of trade unions, workers' groups, students, and peasants. As a 'popular front movement,' it wanted to create a socialist society that emphasizes communal ownership and preserves limited private ownership. It argued that the IMF and international investors created oppressive global conditions. As a result, the MRTA, opposed globalization because

freedom under neo-liberalism is not for people, rather for capital. The function of the state is to be reduced to providing internal and external security. All forms of social policy are to disappear, since they degrade the conditions for capital and create regional disadvantages with respect to the world market ... in Latin America, these policies, with all their catastrophic consequences, are a continuing form of imperialist exploitation.[16]

With a state-centred vision of society that follows the logic of other radical-left groups in Peruvian history, the MRTA opposed globalization because it believed that social progress is best pursued at a domestic level. The MRTA used daring, varied, and populist tactics to send out its message and achieve its goals. In 1988, it kidnapped a retired air force general and distributed a stolen truckload of chickens to striking miners. The two actions aimed to send out a message that the MRTA fights both for people's welfare and against Peruvian elites. The MRTA also frequently used the media to focus public attention on its acts and ideas. In 1985, it created the pirate 'November 4' radio station to broadcast press releases and appeals to boycott elections. In February 1987, it occupied seven Lima radio stations and broadcast an anti-militarist message.

In the 1990s, the MRTA's media tactics targeted the Fujimori government at the same time as the regime's democratic credentials and human rights record were widely criticized in Peru and abroad. While the MRTA consistently charged that Montesino received CIA support, even the U.S. State Department criticized Peru's executive branch for frequently using 'its control of the legislature and the judiciary to the detriment of the democratic process.'[17] A report by a UN Special Rapporteur pinpointed weaknesses in the country's legal system, noting that Fujimori's anti-terrorist legislation failed to meet international standards by

vaguely defining terrorism and treason and by punishing them with disproportionate penalties ... [thereby failing] ... to observe the rule of proportionality. In enacting such measures it failed to abide by its international obligations, and it suspended fundamental rights that are non-derogable even during a state of emergency, principally the right to have an independent and impartial judge to hear one's case.[18]

Reporters Without Borders stated that Fujimori's Peru was a society in which freedom of expression was 'difficult.'[19]

The struggle between the MRTA and the Peruvian state was long and uneven, with both sides experiencing victories and defeats. Facing a fierce government offensive under former President Alan Garcia Perez, the MRTA moved its campaign to the countryside at the end of the 1980s. The government captured MRTA leader Victory Polay in February 1989. On 28 April 1989, the military surrounded a sizable MRTA unit. Heavy fighting and aerial bombing ended in the unit's capture and execution. The MRTA says about sixty-two people, including about twenty civilians, were killed. On 9 January 1990, an MRTA commando shot an official who it said ordered executions: former defence minister E. Lopez Albujar. In 1990, more than 3000 political murders occurred in Peru. In July 1990, Polay and forty-six other guerrillas escaped from Lima's Canto Grande prison through a 315-metre-long tunnel. Polay was re-arrested on 10 June 1992. In April 1992, another MRTA leader, Peter Cardenas Schulte, was captured. In May, police raided an MRTA computer centre and seized important information about the movement's internal structure. Fujimori declared victory over terrorism and the MRTA when thirty Tupac Amaristas were arrested on 30 November 1995 in a plot to occupy the Peruvian Congress and hold it hostage in exchange for jailed militants.

The MRTA stepped into the global mediascape when it seized guests at a reception in honour of the Japanese emperor at that country's ambassadorial residence in Lima on 17 December 1996. The attack once again aimed to force the government to free imprisoned Tupac Amaristas. The group employed sophisticated techniques to explain its acts to a watching world: 'they tried using cellular phones to call TV stations after they took more than 500 hostages in Peru, but the government blocked the signal. They let in 20 photographers, and the government will probably make sure that doesn't happen again.'[20] The MRTA held onto seventy-two people, including President Fujimori's brother, several generals, the heads of police divisions, Peru's foreign minister, supreme court judges, members of congress from the ruling party, and ambassadors from Japan and Bolivia. They remained hostages during a four-month standoff. The MRTA's skillful presentation of its case in the media included a website, which hampered government efforts to control the crisis: '"We can't very well cut phone lines and confiscate computers," says one Peruvian government official.'[21] U.S.-trained special forces finally stormed the embassy on 22 April 1997. All fourteen members of the MRTA 'Commando Edgar Sanchez,' including leader Nestor Cerpa, were killed.

The MRTA Website

The MRTA Solidarity website carried the Tupac Amaru message to the world during the embassy standoff. It was created and managed by the Toronto-based activist group Arm the Spirit (ATS). The server that provided computer space for users was BURN! run by a Latin American solidarity group that accessed the Web through the University of California at San Diego (UCSD). As stated above, the website was eventually removed from the UCSD server since the BURN! website hosted several groups (such as the Revolutionary Armed Forces of Colombia [FARC] and the Kurdistan Worker's Party [PKK]) that, like the MRTA, were considered terrorists by the U.S. government.[22] After passage of the USA Patriot Act, pressure on UCSD forced the removal of many groups from its server.[23] In order to better focus on the variety of Web communications and limit problems inherent in accounting for webpages that often change, this discussion refers to the MRTA Solidarity Page for 22 April 1997 (the day the Peruvian army stormed the compound) as a select case.[24] At the time, the website contained a mix of textual and visual materials, presented documents in several languages, and provided hyperlinks to other like-minded organizations, a variety of media, as well as ATS. The links and the wide variety of materials gave the site an interactive or hypermediated character. Like most small Web activist organizations that have few staff and financial resources and face other barriers to transmitting information on the Internet, the site was not fully synchronous. MRTA statements appeared punctually and were valuable in themselves, but materials on the site did not consistently appear as events occurred. The materials mainly served to comment on events over the preceding days or months.

The main purpose of the website was to supply information about the MRTA, its ideology, and activities. The aim was achieved through several information formats. The first group of documents included regular press releases outlining MRTA views on events. Taking advantage of the medium's variable nature, these were regularly changed. On 22 April 1997, fourteen press releases were available. A second group of texts consisted of interviews with MRTA leaders by both sympathetic newspapers and the mainstream press.[25] A third group of miscellaneous materials included a statement made at the beginning of the hostage crisis, another from an MRTA representative in Europe, a call for peaceful resolution to the Lima incident, a solidarity statement made on International Women's Day, and the text 'Three Months of Occupation! No Surrender and No Defeat!' A final group of more lengthy theoretical-ideological texts that outlined the group's worldview[26] were placed in a separate section to highlight their importance.

Another general heading on the site highlighted the Web's interactive and communicative character by providing solidarity contacts with Latin American and foreign radical organizations that sympathize with the MRTA and its struggle. This heading included a variety of international contacts through regular postal services or 'snail mail,' telephone, and Web media. Various documents supplemented these contacts. One group of documents included texts from the imprisoned American activist Mumia Abu-Jamal, Turkish political prisoners, and the families of German Red Army Faction prisoners.[27] Another group contained weekly news updates about the Lima standoff from the Nicaragua Solidarity Network (NSN), a New York–based organization whose mailing address, telephone number, email address, and website were listed.[28] The NSN uses the Web to maintain contacts with American and international supporters of various Latin American political causes. A regular NSN 'Weekly News Update on the Americas' is available in identical electronic and print versions. The NSN encourages reproduction of its updates and information as well as retransmission of its address.

Significant numbers of solidarity contacts were incorporated into the website through hyperlinks. In addition to English-language connections, the website was linked to sites in German, Italian,[29] Japanese, and Spanish, which enhanced its transnational and communicative character and facilitated global media coverage. Each link provided contacts to international left-wing and Latin American solidarity networks. This multilingual and multinational network endowed the website with its most striking, politically potent, and truly global features, which were augmented by a selection of MRTA press releases posted in Spanish, English, German, and Danish. The site's multilingualism showed how non-state actors push the boundaries of IT use even though, in 1997, the Web's dominant language was English. The multiple languages on the website demonstrated how Web media facilitate exchange of information not available in a major international language. Danish materials were included due to the group's significant support community in that country, which was the location of the official MRTA page.[30]

The variety of languages also illustrated how Web communication is more highly focused than traditional media: the inclusion of Italian-, German-, and Danish-language materials reflected specific MRTA support demographics rather than market imperatives. A final element in the links section was a short video clip of MRTA guerrillas preparing their departure for the Japanese embassy seizure. The quality of the video images was mediocre and revealed nothing critical about the MRTA, but showed how a group with limited means could hypermediate, represent itself to a global public, and circumvent government- or corporate-generated images. In fact, the site also archived government and corporate print news media. One section assembled Peruvian and international press coverage of the hostage crisis that was organized into groups for December 1996, January to April 1997. It also included the Lima centre-left newspaper *La República*. The archive demonstrated that the Web is a hypermediated environment in which non-state actors have potential input into mainstream information sources.

In principle, Web activists aim to send messages to as many people as possible. In fact, they also need to target a receptive public. MRTA messages went to a public shaped by the limits of Internet access. The webmaster and users could monitor the number of 'hits' (visitors) to the site by means of a website counter. The number of hits suggested the shape of the MRTA 'public.' In April 1997, *The Wall Street Journal* reported that the site had become 'a hot spot in cyberspace. Internet surfers have logged on more than 16,000 times to one site set up by Tupac Amaru sympathizers in the U.S. and Canada 10 days ago.'[31] By 22 April 1997, the site had been visited 88,035 times. In the period between the seizure of the embassy on 17 December 1996 and 22 April 1997 (127 days), the site had 693.18 hits per day. From 22 April to 7 July 1997, the number of hits increased to 708.87 per day. They sharply decreased afterward: only 137.53 hits per day between 7 July and 19 July 1997. By 7 July 1997, the total number of hits reached 142,618 and by 19 July, 144,406. By 13 May 1998, 172,620 hits were recorded, increasing by 28,214 or just over 100 hits per day in a nine-month period.[32]

As in the case of RAWA, the figures on the number of website hits indicate that a message did reach a global public and that offline factors shaped its reception. In particular, the intense media focus on the embassy standoff appears to have stimulated the number of hits. A later loss of media attention led to fewer hits in the following weeks and months. The decline suggests that Web communication is linked to other media since increased attention by traditional outlets stimulated interest in the site. An important analytical question raised by this relationship is the connection, both network and conceptual, between the

Web and other media. While MRTA Web activists used the media to respond to their particular circumstances, such use embodied a specific global network that they established with other non-state actors and illustrated how the Web is an assembly of pre-existing media that feed into its information flows as determined by events at specific times. In the case of the MRTA, widespread use of the site by journalists was a factor that certainly affected the number of hits. Their recourse to the site gave the MRTA input into the global mediascape but also subjected their message to the influence of other global actors, in particular global media corporations that are often criticized for 'homogenizing' issues.

The relation between Web activism and the global mediascape is undoubtedly highly complex. By employing the Web, both traditional media and activists communicate more easily and frequently than possible through radio, television, or newspapers. If many hits on the MRTA website were indeed from journalists and not the general public, the Web's limits and potential are clear because it touches a restrained but influential minority of the global population. The Web enhances communication capacities for users with suitable skills, income, access, or, ironically, locations (such as in Western societies). Viewed from this angle, Web activism does not so much short-circuit as circumvent and transform previous hierarchies, creating new global elites that access, produce, and transmit specific types of information.

Insurgency Online as Media Relay

Before the removal of the English-language version from the Web, the MRTA website was a rich resource for analysts of politically violent non-state actors in a context of globalization and rapid IT development in which research is difficult and controversial. Comprehensive analysis of such groups is always hampered by lack of information about given conditions, vague categories applied across vast cultural and political divides, implicit assumptions about the nature of political violence, as well as the medium's rapidly changing character and content. Rather than conceiving violence as an unfortunate and recurrent feature of most political systems, many analyses operate from the Westphalian or state-centred mind-set in which all non-state violence is considered illegitimate and all governments are assumed to be legitimate, based in popular consent and rule of law, and facilitating interest representation by various social groups. However, the image of a legitimate state securing citizen rights and interests is contradicted by the fragile legal frameworks, ongoing political repression, and information manipulation that is routine in many states.[33] Where peaceful opposition is impossible or futile, violence is a tool that draws attention to opposition views or makes demands on unresponsive systems. This point is especially critical in a post-9/11 world in which, despite U.S. efforts to dichotomize between the supporters and opponents of 'terrorism,' global conflict is asymmetrical, highly fragmented, and complex.

The most common representation of non-state political violence in Western media (newspapers, radio, television) is strongly influenced by a Westphalian perspective. British and American media often treat non-state political violence as a symptom of individual psychological problems that is condemned rather than interpreted. When the MRTA seized the Japanese embassy in Lima in December 1996, one U.S. media report quickly linked the event to anxieties over the allegedly dangerous and uncontrollable consequences of IT, stating that 'like many other radical or revolutionary groups in the developing world, the Tupac Amaru Revolutionary Movement, or MRTA, has found allies in cyberspace.'[34] Like all political movements, the MRTA was certainly producing propaganda. Indeed, their non-state political violence was 'propaganda by deed' that aimed to build morale, advertise, disorient, physically eliminate foes, and provoke government reaction.[35] The combined physical and non-physical impacts of MRTA acts imbued their violent tactics with a complexity and variety that parallel conventional political behaviour. This means that the MRTA need to be interpreted in a sophisticated manner. To assess the meaning of violence in specific circumstances, ample contextual information is needed to account for the 'subjective' perspective that constitutes the social knowledge and values that lie behind behaviour.[36] In this sense, analysing political violence is 'more like interpreting a constellation of symptoms than tracing a chain of causes' since all explanation is partial and comprehensive significance is difficult to understand and express.[37] By analysing Web activism, we get a partial view of the MRTA, but one that is no less partial than, say, analyses of 'terrorism.'

The conceptual barriers to analysis are compounded by other factors. A major barrier is the analytic use of the term 'terrorism.'[38] The discipline of IR is unfortunately marked by over-conceptualization and lack of case fieldwork, which compound these barriers. This problem is illustrated by the three groups in this study, all of which exist in contexts marked by complete or partial failure of state formation. In one case (Northern Ireland), this weakness is present within the territory of a state (the UK) that is often considered to epitomize the Westphalian state. The resulting studies of terrorism make sweeping generalizations about terrorism and political violence on the basis of one or a very few examples. My own work on terrorism focused on a group (*Action directe*) that was very dangerous to a specific (military, government, and business) elite within a particular society (France), but actually more threatened than threatening in the final analysis. The case contradicted the image of a serious menace to the state or social order.[39]

These interpretative problems are compounded when 'terrorism' is simplistically transposed into an electronic format. 'Terrorism' elicits an emotional and moral response even when used in analytical contexts. The term triggers the assumption of a serious threat to physical integrity in real social settings and to the integrity of systems in an electronic context. However, the various viruses and website defacement that regularly appear as information terrorism are ultimately more of a nuisance factor than an overwhelming threat to systems security. Considered in this light, 'terrorism' misleads. Violent acts by secret or semi-secret non-state actors aim to gain maximum effect from surprise and little advance warning. Analysis may not have access to information that links terrorists to a community of interests, needs, and values that would help explain the purposes behind acts. In addition, globalization and IT structure a post-territorial environment in which more fluid political identities are established and maintained across borders, continents, and time zones. Where politics once grew from ties built in meetings, signed agreements, physical proximity, and local interaction, a rapidly changing network of international 'non-place' communities now exists through fax machines, personal data assistants (PDAs), satellite and cable television, text messaging, video games, virtual reality, telephones, email, the Web, blogging, and computers. In this diverse and globalized setting, providing information per se is not 'terrorism' even if transmitted in support of politically violent non-state actors. Support and encouragement via IT might aid violence, but they are not inseparable from it.

The impact of the globalized and globalizing Web, like that of television in the 1940s, is unclear, but has several discernible features. First, the Web facilitates direct contact between users in non-contiguous physical locations. Senders and receivers enter two-way interactive contact. Direct exchange occurs alongside the channelling of passive users through centralized outlets in traditional media.[40] Second, Web synchronicity allows users to log on at any time and receive information from distant physical locations as events occur. Third, Web technology hypermediates; that is, it integrates radio, television, telephone, email, multi-user chat rooms and conferencing, and other functions while providing access to powerful transnational computer databases. Fourth, the Web has a network transmission structure that makes limiting, blocking, censoring, or deflecting messages quite difficult, which facilitates non-linear communication and transcends physical-place hierarchies. Fifth, Web media are *variable* forms of communication in which texts as well as audio, video, and other materials are transmitted or altered to suit needs. Finally, Web communication reaches a dynamic though limited demographic base. Access is shaped by specific class, race, ethnic, language, age, and educational groups,[41] but changes constantly and rapidly. At the time of the Peruvian hostage crisis in 1996–7, Web use was limited to those with higher incomes, institutional affiliation, textual literacy, higher education, and some informal or formal computer-based education. Even U.S. users

> require a more sophisticated and expensive network connection, have a stronger gender, education, and income bias and are more likely to be students. In addition to the cost factor, there are also differences in how much people value CMC [computer-mediated communication] use. Younger men typically have a higher interest in technology, increasing the perceived benefit of CMC use and reducing the net cost.[42]

Despite the rhetoric of technology-driven forms of globalization, the transformation brought on by the Web is highly relative because access is very uneven for regions and sectors of the global population (ranging from next to no access in Africa to high levels in North America, Japan, and Western Europe). These limits can even be seen in the United States, a society in which the Internet is widely used but where barriers to access and use still persist. In the United States, Katz and Rice identified

> three key barriers to Internet usage – cost, access, and complexity. Two of these – cost and access – were more strongly felt by nonusers, perhaps reflecting their lower incomes (ability to pay for the Internet) and educational achievements (ability to navigate the Internet). Most significantly,

both users and nonusers were equally concerned about Internet complexity. Without improvements here, frustration levels will remain high, and potential user benefits will in many cases go unrealized.[43]

Outside the United States, access issues are even more striking and often daunting. A 2001 Organisation for Economic Co-operation and Development (OECD) document noted that

> In 1997, more than 30 African countries had less than one telephone line per 100 people, according to OECD figures. It is not simply that the 'haves' are at an advantage, but that the 'have-nots' are at increasing risk of social and economic exclusion. Countries which lack a firm ICT infrastructure become marginalised as electronic commerce grows in importance. They are incapable of sharing in the new route to prosperity which e-commerce affords, and remain dependent on the export of basic commodities, for which the world price is often in decline. Africa's share of world trade has fallen from about 4% in 1980 to less than 2% today, according to IMF figures.[44]

In spite of the specific demographics of the Internet, flexible and relatively cheap Web communication still gives non-state actors an advantage in reaching a once-inaccessible public. The International Telecommunications Union (ITU) estimated in 2003 that there were 665 million Internet users in the world. Yet, while the Web more easily transcends hierarchies based on language, editorial content, and dialogue than do radio, television, or print media, these hierarchies are transformed, not eliminated. There was a global average of 1027.92 Internet users per 10,000 people in 2002. As noted in the Introduction, these users are spread unevenly around the world, being concentrated among certain classes, regions, and other groups

Toward a Theory of Web Activism

MRTA Web activism was very successful in providing information about the organization's goals and activities to global civil society. A major issue raised by their Web activism is the desirability, possibility, and feasibility of governments blocking information. Advocating armed resistance to democracies that facilitate free markets, social justice, and open public debate is problematic, but as the U.S. State

Department and other organizations note, Fujimori's Peru did not correspond to the democratic model. Some states object to Web activism that draws external opinion and support into domestic politics. Their unease is understandable, but in the case of Fujimori's regime, the effective alternative was to encourage political and social equality, rule of law, and open debate instead of censorship. Overall, social, economic, and political conditions are more of a threat to states than is Web activism; democratic societies have little to fear. More than violating sovereignty, Web activism tests the limits of free expression and irritates governments that do not foster democratic conditions. In spite of the so-called IT revolution, it is by no means clear that international communication is a decisive lever of power in an age in which the United States, in spite of a wealth of information that suggested the risks, floundered into an unprovoked war in Iraq.

Web activism has moved marginal non-state actors into greater public awareness. It also resembles Web communication by conventional groups, although it has different ends. The goals of marginal non-state actors are to place unusual, unpopular, and/or uncommon ideas before a public on the Web. The ends of mainstream organizations are many but do *not* include spreading the ideas of unusual, unpopular, and/or uncommon organizations. Until recently, obtaining information about extremist, radical, or illegal opposition groups was difficult even for specialists.[45] The Web provides extra-legal non-state actors with relatively easy access to a wider public, a venue for publicity, and a 'marketplace for ideas.' As such, Web activism is part of a global boundary crisis that is altering the conditions for politics and identities. The Web is a superb public relations tool, a means to tell an activist story, and a way to avoid regulation and control. It enhances non-state actors' ability to locate and communicate with like-minded people. A small organization can appear large, well-financed, and well-organized, and fabricate a sophisticated representation of its legitimacy.

The potential overall results of Web activism might include stronger self-identification, recruitment, and morale building. The MRTA website provided access to the global mediascape, allowed better communication with other groups, and even drew its ideological enemies closer: Shining Path's page can be used to access the Web page of its MRTA rival.[46] At the same time, the MRTA also provoked an online Peruvian government response. An 'Association for Truth' (Aprodev) put up a 'blacklist' of Peruvians on its website in November 1998. These individuals were seven journalists who had questioned government actions or the methods of the SIN.[47] The site presented threats and accusations from the tabloid press as well as very detailed biographies that mixed true and false information. The seven journalists filed a complaint against Hector Faisal, Aprodev's legal representative, in April 1999. The magistrates in the investigation declared the complaint inadmissible one month later. In October 1999, Faisal was acquitted. However, in January 2001, a few weeks after Fujimori was deposed, a magistrate forbade Faisal from leaving the country and his possessions were impounded.

How long non-state actors will access a wider public through the Web is unclear. Web activism flourishes due to fragmentation, erosion of national boundaries, increased cynicism about governments, and awareness of transnational issues such as the environment and human rights. These conditions might be short-term spinoff effects of globalization rather than long-term transformations. Limited Web access in most regions of the world suggests that broader political participation is probably not at hand. Computer literacy, network access, educational levels, and economic resources would have to increase many times over current levels. Beyond this, other limits exist. The explosion of online information requires time, energy, and mental effort to properly sort, filter, interpret, and utilize it. The results of IT are not a priori known. In some cases, citizens have been empowered (as in the Philippine cell phones used to oppose the Marcos government, and Iranian audiocassettes used to transmit illegal information), while in others states have benefited (in China and Saudi Arabia, Web technology has been used for state development ends). This has led observers such as Baudrillard to criticize today's IT as essentially a homogenizing force. Baudrillard's comments echo Baudelaire's condemnation of an earlier globalizing IT, photography. Baudelaire in his time, like Baudrillard in ours, was concerned that the notion that photography could accurately and objectively represent reality would 'confirm fools in their faith' and lead to a narrowing view of society and human relations.[48]

Web activism does change the conditions of sovereignty, but in ways that are unexpected. By exposing the corruption, incompetence, and human rights abuses of the Fujimori government, MRTA Web activism drove another nail into what eventually became that regime's coffin. That this was achieved in part through the mobilization of global civil society and the assistance of activists in California and Toronto illustrates the transnational impact of the Web. The story does not end here. By August 2003, the MRTA Solidarity Page was unavailable in English, the language in which it created a sensation just six short years before. Pressure from the Bush administration and the legislative impact of The USA Patriot Act mentioned above in fact led to not only the censoring but the dismantling of the English-language MRTA website. While the transformed conditions of sovereignty can bring forth new information from non-state actors, they can also provoke a regulatory push from powerful states. This complex push and pull between states and non-state actors is prominent in global politics, where states once had unquestioned dominance. Much of the Bush administration's actions regarding Web activism such as the MRTA seems to confirm a pessimistic assessment. At the same time, the circulation of photographs of prisoner abuse by U.S. military personnel at Abu Ghraib prison in 2003 shows how IT continues to have unanticipated impacts.

5 Conclusion: Web Activism – A Messenger That Shapes Perceptions

Bertolt Brecht once stated that 'the radio would be the finest possible communication apparatus in public life ... if it knew how to receive as well as transmit, how to let the listener speak as well as hear, how to bring him into a relationship instead of isolating him.'[1] Brecht recognized that two-way and multiple-user communication devices could transform society. Today, such devices are present in the form of desktop, laptop, and hand-held devices in many settings: integrating pre-existing technologies such as computers and photography on a single platform; introducing new values, ideas, and interests in many contexts; privatizing information about world events for a significant minority of the global population; and introducing a range of non-state actors who have global communication abilities into the global-scape. As Brecht foresaw in relation to radio, communication devices promise to foster a range of new human relations.

In 1936, several years after Brecht's musing, Walter Benjamin's 'The Work of Art in the Age of Mechanical Reproduction' systematically analysed the impact of industrial production on representation. Benjamin argued that mechanical reproduction alters artistic expression by removing it from a context of ritual and tradition, which he said destroys the 'aura' of art. For Benjamin, this transformation endows representation with a political-revolutionary potential that is both empowering and dangerous: 'for the first time in world history, mechanical reproduction emancipates the work of art from its parasitical dependence on ritual [and] the total function of art is reversed. Instead of being based on ritual, it begins to be based on another practice – politics.'[2] Benjamin argued that art in the mechanical age has the potential to move closer to the lived experience of industrial workers by consciously referring to a political universe of representation rather than posing as the embodiment of a sacred moral-emotional order. The intrusion of politics into art confirmed Benjamin's view that representation is historical and malleable. He believed that such art might immediately and authentically link human beings together by directly entering their life and work instead of remaining the preserve of an elite. In Benjamin's view, mechanical reproduction could democratize culture by awakening artistic expression to historical meaning. Seventy years later, the Web promises to realize this potential through its accessibility, variability, and multi-directional communication.

Benjamin also warned that technology's promise is mitigated: 'the destructiveness of war furnishes proof that society has not been mature enough to incorporate technology as its organ, that technology has not been sufficiently developed to cope with the elemental forces of society.'[3] His warning has been substantiated by events. IT has abolished distance between subjects, but has also facilitated ethnic cleansings, eugenics, and social engineering. In 1994, the Rwandan genocide was supported and encouraged by radio broadcasts that sent instructions to kill. The promise of communication became the end of communication as radio was used as a tool to break, not construct, communities.[4] A misplaced longing to reconstruct or reconstitute mythic perfection and innocence as a 'nation' turned Rwandan radio into a springboard for mass murder. Beyond the gateways and channels ringed by encoded fortifications, then, IT carries a paradox that binds the Web to

> an even more radical divide between those who will live under the empire of real time essential to their economic activities at the heart of the virtual community of the *world city*, and those, most destitute than ever, who will survive in the real space of *local towns*, that great planetary wasteland that will in the future, bring together the only too real community of those who no longer have a job or a place to live that are likely to promote harmonious and lasting socialization.[5]

As Brecht anticipated, devices now spread previously inaccessible information and significantly transform political communication by allowing non-state actors to directly intervene in the global mediascape. For the IT regulated by states after the creation of the ITU in 1865 (telegraphy, telephones, radio, television, etc.), the transformation has had unimagined impact as new contents, issues, and movements entered the parameters of global politics and transnational security. Web activism is one example in which non-state actors alter global power by reshaping perceptions in a mediascape that was formerly a monopoly for the state.

The links between IT applications and contemporary security are apparent in a period in which communication devices support and encourage human interaction. The wide variety of today's IT applications function in a multi-centric world in which non-state actors are players, but not the dominant ones. One area of IT applications is the military sector, which is based in states that still control coercive force on a global scale. Military applications of IT can be seen in satellite images, the use of cell phones by the U.S. military to contact Iraqi commanders in the 2003 invasion, air-borne drones, night vision and sensor devices, and other areas. IT is also relevant to global power in the context of cybersecurity, an area that is most associated with public anxieties over new technologies. The D DOS attacks on the website of the Qatari-based Al Jazeera cable network mentioned in chapter one are an example of this application.[6] A third area in which IT relates to global power, Web activism, introduces a struggle over perceptions and reshaping of the boundaries of identity and politics into global security. This area concerns the global mediascape that states previously regulated and controlled.

Websites and Boundary Shifts

Web activism is electronic direct action by means of multi-directional global communications devices. It is a form of political behaviour that uses the Web to shape identity and perceptions, and transform the relations that underlie global politics. Graham Meikle's *Future Active* points to a variety of Web activists who extend offline struggles to online environments by 'raising awareness of the issues concerned, and this means getting more coverage than the purely online.'[7] Meikle examines anti-globalization activists, the Serb opposition radio station B-92, the Electronic Disturbance Theater, hacktivism,[8] and alternative and independent media. Similarly, the MRTA, RAWA, and IRSM each reflect variety, produce distinct analytical results, and use text and images to represent a specific political vision. The results are not revolutionary in a conventional political sense. Instead, Web activism substantiates the model of a multilayered and fragmented global mediascape in which boundaries and identities are on the move and where representation, at a time in which images of Osama bin Laden shifted American public opinion just days before the 2004 presidential election, has become a real security concern.

Multimedia Web Activism

In the face of rising website censorship, the contours of Web activism have changed as a function of its flexible technological environment. Web use that is more anonymous or more difficult to locate, such as the transfer of multimedia files, is widely practised.[9] The Web carries many messages from groups that support Al Qaeda's view that the West must be opposed by violence to prevent attacks on Islam. The formats of these messages show that the Web can be adapted for covert non-state actors as websites that support radical groups are shut down or limited. Islamic extremists circumvent censorship or easy observation by using transferable files. These files are not as readily accessible as websites since they are emailed between recipients or accessed by clicking on webpage icons.

A striking example of extremist Islamic multimedia Web activism is a Shockwave document[10] found on the website of British-based firm CityLink Computers[11] in July 2003. The file is an audio recording in Arabic of a *du'a* (prayer) by Sheikh Muhammed Al Mohaisany that was delivered at the Masjid Al Haram (Grand Mosque) in Mecca during Ramadan in 2001.[12] It is accompanied by subtitles in English and a powerful selection of digital photographs. The *du'a* is an arguably unique stand-alone document that contains no claim of authorship.[13] The text says that the Saudi Arabian government arrested Al Mohaisany immediately after he delivered the prayer.[14] In his prayer, Al Mohaisany asks for victory for the 'Mujahideen,' complains of 'the injustice of the spiteful Christians,' calls on God to direct His 'forces against America,' charging that it has 'killed Your slaves' and 'insulted Your religion,' and invokes hurricanes as 'a constant for them' (i.e., Americans). The prayer further asks for release of 'our captured brothers' (i.e., prisoners in Guantanamo Bay), eradication of those who torture them (U.S. soldiers), saving 'Al-Aqsa from the cruelty of the Jews,' and protection for 'the hard-working scholars' (a passage accompanied by a photo of Osama bin Laden).

The *du'a* file juxtaposes audio recording and text with digital images. The images include long shots of Mecca, the Al-Aqsa Mosque, Osama bin Laden, Chechen fighters, suicide bombers, the Palestinian Intifada, destruction of the World Trade Center, Israeli and American soldiers, the burning of U.S. flags, and a downed U.S. helicopter. Other images show dead Muslim children in Palestine and Afghanistan, U.S. prisoners in the 2003 Iraq conflict, captured Islamic activists, Camp Delta in Guantanamo Bay, transport of Taliban prisoners in U.S. aircraft, scenes of destruction in Jenin, Israeli troops near Al-Aqsa mosque, and various political leaders, including Saudi monarch King Fahd, Palestinian leader Yasser Arafat, Pakistani president Pervez Musharraf, U.S. president George Bush, British prime minister Tony Blair, U.S. secretary of state Colin Powell, Egyptian president Hosni Mubarak, and former U.S. president Bill Clinton.

Like the other forms of Web activism discussed here, the transfer of multimedia files has implications for IR. Although the cultural dimension of IT is still largely ignored or sidelined as 'propaganda' in IR, the rise of the Web means that the private transformation of perceptions is a daily experience for a significant and influential minority of the global population. As such, Web activism shapes the world in which we live, blurs conventional distinctions, and alters the parameters of state action as well as group and individual identities. In the case of the *du'a*, the innovative dimension of communication devices is highlighted in terms that recall Benjamin's suggestion that 'mute' photography needs to be supplemented by captions, which is exactly what the multimedia file does as a matter of course.[15] Multimedia Web activism influences perceptions and boundaries through images that have a strong emotional-moral impact and a text that moulds the contents of that impact.

The *du'a* of Sheikh Al Mohaisany further exemplifies the challenge that the Web poses to conventional identity and state boundaries. To be sure, transnational ideological radicalism is not new. The precedents for contemporary transnational Islamic extremism were the global communist and fascist movements of the twentieth century insofar as the latter crossed state boundaries and were ideological in character. Of course, unlike today's Islamic extremists, the latter were not religious movements and did not benefit from the extensity, intensity, or velocity of modern IT or, for that matter, draw advantage from a global scene in which a variety of factors (from AIDS-SARS-Ebola to migration to stock markets and finance to technological exchange) mitigate the centrality of the state. The *du'a* appeared in a context of heterogeneous and overlapping political forms, organizations, and expressions, a setting in which IT contribute[s] to continuity in the areas of international security and in the world economy by enhancing the capabilities of states while at the same time they seem to be an element for change in these traditional areas by enhancing the capabilities of actors in the multicentric world.[16]

Influence, value, and participation now have a relative link to hierarchical organization, territorial continuity, and spatiality. Multimedia Web activism is a political behaviour that expresses the transformation. The power of the *du'a* is based in a presentation of photographs, audio text in Arabic, and English subtitles that aim directly at viewers' perceptions. For an English-speaking viewer, positioning photographs with text or caption[17] has a powerful impact. The overlap plays with perceptions by blurring boundaries and identities in the absence of a prevailing meta-narrative. In this way, the file realizes Benjamin's and Sontag's interpretations of photography: a mute, contentless image is captioned. The recording of Al Mohaisany's *du'a* is a performative form that transforms an idiosyncratic interpretation of Islam into a powerful transgression of the boundaries and identities of global security.

IT in the Global-scape: Innovation, Continuity, and Perception

Since IT supported and encouraged identity and state formation in the nineteenth and twentieth centuries, the extent and intensity of the Web suggest its impacts might be as far-reaching. Since the Internet platform mixes technological practices, its paradoxical impacts on international affairs are still incompletely apprehended. Attitudes toward the 'truth-values' of the Web will change as societies and individuals are more fully able to assess its impacts and limits. However, one way to evaluate the Web's purportedly authentic and immediate representation of individual and social realities is by reference to older technological practices such as photography, computing, TV, or radio. Widely used and highly visible on the Web, photography is a controversial practice in which 'the press tends to play down the way it uses digital imaging techniques in photojournalism, since it recognizes that electronic retouching of imagery destroys the documentary value of witnessing.'[18] The text-based IT seen on the Web elicits different responses from a cultural mind-set that values written expression of scientific, spiritual, and intellectual experiences.[19] The mind-set devalues images and places words in a privileged position near a purported truth. It is also present when email and website text are protected from interference by hackers and photos are assumed to be inherently unstable indicators of value. Given this bias, the Web's impact obviously varies. While

> the popular notion has it that as images became increasingly sophisticated, their power grew ... the opposite may be the case. As images become more complex and multimediated, their truth-value is communicated in configurations that allow us to see less: in some cases it dissipates; in others it is reconfigured; in still others it completely disappears. Proclaimed vehicles by which we bear witness to events of the present and past, images in some cases become deceptively ambivalent, communicating contradictory messages about their ability to replicate slices of reality and their ability to aggressively reconstruct it, often to the point of fabrication.[20]

For the Web to provide lower thresholds of conflict, more effective economic practices, and greater inter-state cooperation by means of immediate and authentic communication, this bias would need to be overcome. However, cultural ambivalence toward photographic practices means that the Web's impact might not be straightforward. Like TV, the impact of the Web depends on a constellation of factors. This ambivalence could even introduce a beneficial element of doubt into the certainties of statecraft, war, and IR in general. Heated debate over alleged weapons of mass destruction in Iraq did not prevent war in 2003, but widespread information on the issue cast serious doubt on simplistic labels such as 'threats' and 'enemies.'

Western conflict and security practices were defined by the Westphalian structures that informed Louis XIV and his military architect Sébastien le Prestre de Vauban. Having inherited an exposed and fractious kingdom, the *Roi soleil* set out to recast the mandate of heaven in his person and manifest it as the *grandeur* of France in a suitable and viable embodiment. Vauban designed a series of defensible fortifications that traced the general borders of modern France. Many of these structures can still be seen, for example, near Belgium or southwest of the town of Collioure in the Pyrénées-Orientales, a part of Catalonia wrenched away from Spain in the seventeenth century.[21] Stitching together a realm with stone fortifications and defendable borders created the French *hexagone*. The Sun King's craft is a key component of the spatial conception of power that, although a peculiar specialty of the Western group of nations, has been exported and taken up with gusto in India, Indonesia, and elsewhere.[22]

The creation of France shows how the principle of territoriality became operational. IR and much world history since the seventeenth century are marked by the export of such techniques to the rest of the world through imperialism and its intended and unintended consequences.[23] As this example shows, boundary definition is an old theme in IR. It has also accompanied conflict since at least the nineteenth century, when the telegraph was used to facilitate European imperialism and consolidate global dominance. Seen as such, Web communication is part of a bundle of globalizing practices that also include military organization and legal codification. These practices are embedded in the structure of global power along with their contradictory implications (primarily seen as 'underdevelopment' of certain world regions and 'advanced' status in others).

The impact of today's IT is paradoxical. Overwhelming U.S. economic, technological, and military assets have not resolved conflict and have failed to definitively structure perceptions. The U.S. idea of power in the Bush years is still linked to a view of politics as at some level an ability to be physically present, occupy space, and define a place. This Roman idea of greatness is linked to the amount of territory occupied, but expresses the interaction between peoples at a distance who locate one another on maps; 'for centuries, our knowledge of faraway peoples and places depended on reports and maps from courageous (and sometimes foolhardy) sailors.'[24] As an electronic means of communication, the Web seems a priori facultative since it is 'flat'; that is, it does not encompass the depth of human relations, only their breadth (in Katz and Rice's terms, it 'supports and encourages'). To explore the impact of the Web thus entails reconsidering the idea of power.[25]

Web Activism, Conflict, and Security

The power of non-state actors as well as the multitude of movements on the Web illustrate the complexity of any discussion of the global mediascape. These groups range from terrorist organizations such as Al Qaeda to the Raëlian movement, whose claim to have cloned human beings usurps another traditional area of state control: management of procreation.[26] The various tendencies in the global-scape mean that it is accurate to speak of a post-realist world in which the mediascape and power are no longer a state prerogative. It is not clear that in spite of enormous technological, economic, and military advance over all other players the United States is a hegemonic power that directs the internal affairs of other states, controls international alliances, and dictates international economic arrangements. The limits on the single most powerful global actor underline the uneven capacities of all states. Paradoxically, states remain the single most important agents of international power.

The post-realist conditions in Afghanistan and the Palestine–Israel conflict allow powerful states to surgically displace and marginalize opponents, but they cannot end chronic violence or adequately address underlying social grievances. In this setting, 9/11 attested to the global reach of organized violence by non-state actors, a qualitative shift that further affected the centrality of states in world affairs:

> The contours of these 'new wars' are distinctive in many respects because the range of social and political groups involved no longer fit the pattern of a classical interstate war; the type of violence deployed by the terrorist aggressors is no longer carried out by the agents of a state (although states, or parts of states, may have a supporting role); violence is dispersed, fragmented and directed against citizens; and political aims are combined with the deliberate commission of atrocities which are a massive violation of human rights.[27]

Global organized violence by non-state actors, whether in attacks on the USS *Cole* and the U.S. Embassy in Kenya or in Bali, Indonesia, and Mombassa, Kenya, has eclipsed the management rhetoric of the 1990s. It has been replaced by apocalyptic visions and language: U.S. president George Bush refers to 'evildoers' and the 'axis of evil' in speeches; Israelis speak of the survival of a Jewish state to rationalize human rights violations in the West Bank and Gaza; Palestinian and Chechen populations tolerate strategies of suicide bombings; and Al Qaeda continues a methodical and murderous program designed to discredit the West. All the while, states have less control over the mediascape that conveys messages and shapes perceptions.

In the contemporary mediascape, multimedia files and website communication by non-state actors alter the boundaries of political society and identities. Both types of Web activism are image- and text-based representations that transgress identity, space, and the legitimation capacities of states like the United States in different manners: websites tend to *inform* while multimedia files *portray.* By combining text and images, websites offer allegedly immediate and authentic information from specific non-state actors. Multimedia more directly addresses the viewer's emotions and morality. When artifacts like the *du'a* express a sense of grievance and injustice rather than set out a program, Web activism's symbolic, image-driven, and identarian features upstage its information aspects. Multimedia artifacts' power is an ability to present a coherent vision of reality that blends words and images.

The enhanced representation and post-territoriality of Web activism raise the issue of characterizing a shift across a range of phenomena. Do changed technologies of representation signal a deep (fundamental) or extensive (wide-ranging) transformation? Will the shift produce new elites and new rigidities? The cases of RAWA, the MRTA, and IRSM suggest that the Web transforms global politics in a way and to an extent that is more modest than was apparent even a few years ago. Web activists *have* an impact, but they do not overthrow states or necessarily even redirect public policies. The change is wide-ranging rather than deep. It has occurred globally but has not resulted in conventional political change (i.e., new governments or new constitutions). Instead of a tool for revolutionary transformation, Web activism is a powerful method for political organizations of all stripes in precise circumstances that favour their particular messages.

'Media are contentless' – A Messenger That Shapes Perceptions

Media are not merely messengers: they are also messengers that shape perceptions in distinct ways depending on what IT is used. Since Web technology integrates previous communication practices, it is logical to integrate evaluations of those IT into an assessment of the global mediascape. Sontag's writing on photography says that photographic images are mute and have 'multiple meanings; indeed, to see something in the form of a photograph is to encounter a potential object of fascination.'[28] Paul Dourish argues that computers have a specific impact on how humans use them since their design 'favours performance over convenience, and places a premium on the computer's time rather than people's time.'[29] Both point to the communicative limits of media and the technical constraints of devices, which are in contrast with widespread unease over the allegedly limitless possibilities and boundless threats of technological innovation.

IR needs to develop theoretical frameworks adapted to specific IT, investigate the types of groups that use specific applications, and assess the impact of use across a selection of cases. Researchers must learn to 'read,' describe, and interpret the political content in IT applications as they learned to read newspapers, listen to the radio, and watch TV after the initial thrill of communicativeness passed.[30] The MRTA, RAWA, and the IRSM operate with varying degrees and measures of success in a complex mediascape. They have little in common beyond the use of the Web to send messages to global civil society, a struggle against very different opponents, and the failure to produce conventional institutional and governmental results. Each example embodies the Web's potential impact by transgressing, re-articulating, and reshaping the boundaries of political societies and identities. This potential is all the more real in a global-scape marked by variable layers of authority, action, and power, which Rosenau calls a 'bifurcation' between state-centric and multi-centric forms of international action.[31]

Web activism shows one impact of an Internet platform that draws together various existing technological practices. This analysis has unpacked some features of specific Web activism in order to integrate insights regarding their impact with a broader discussion of IT and IR. A path for continued analysis is examination of the component applications for the platform, such as photography, television, radio, and computers as well as practices such as graphic design. Photographic practice has traditionally been divided between those who argue that photographers should intrude into the world and others who seek to efface their presence from photographs. In the Western tradition of representation, this notion of self-effacement posits images as 'authentic' and 'accurate' views of reality. In the film *Nights of Cabiria*, a Roman prostitute dreams of romance and respectability. Betrayed and robbed by her tricks, the Giulietta Masina character nonetheless loves, lives, rebels, is moral, is human. She turns to the viewer, eyes brimming with tears. We understand, seem to experience, her will to live and love, understand that this is a story for sentimentalists. 'It's only a story,' her eyes say to us. A moment becomes a documentary of the life of a Roman whore, becomes us, and we are freed to live pain, cry, suffer, shed the Cartesian strait-jacket. Masina's knowing look aligns itself with Warhol's cheeky Marilyn as reproduction after reproduction stretches to the horizon.

Self-effacement in photo-magazines such as *Paris Match* and *Life* purportedly portrays 'daily life' at specific times, presents a record of events, and provides photo-documentation. Photographs mark particular historical events or periods. Self-effacement raises the question of the objectivity and the status of images. Photo-narration has long played a role in conflict by fostering support for military campaigns.[32] On the Web, photographs and video clips enter a medium that purportedly presents a range of values, ideas, causes, and situations. The context of Web photography is a social formation called the information society that, like others before it, is marked by inequities, interests, elites, and values, but claims to present a 'global landscape.' The problem, as one observer notes in relation to satellite television, is that 'while "global village"-inspired rhetoric touted the utopian promise of new satellite technology, it was complicit with Western discourses of development that worked to subjugate non-Western and postcolonial cultures and peoples.'[33] IT thus provides benefits, but also risks favouring some groups or regions over others both in how they operate and in the resulting perceptions of the world.

Globalization is a cultural, economic, technological, social, demographic, and military process that brings new groups, ideas, values, interests, and forces into a global-scape of contact and conflict.[34] Web activism is one site where values are introduced, debated, and contested. While Brecht hoped that radio (and other media) would foster authentic human relations, our lived experience of politics is still marked by images, copies, representations, and appearances. 'Reality' and 'image' have not been neatly separated. And images can overtake reality. As a United States reeling from the impact of 9/11 stumbled into a profoundly consequential Iraqi conflict in 2003, an obsessive focus on the image of terrorism contaminated the highest levels of defence and foreign policy thought and planning. The results are a focus on the so-called 'axis of evil,' failure to capture Al Qaeda mastermind bin Laden and Taliban leader Mohammed Omar, continuing chaos and insecurity in Afghanistan, explosive violence in the Middle East, and the Iraqi quagmire. The danger is that security challenges will be met in purely military terms while conflict is more multi-layered, innovative, representational, perceptual, and image-based than ever given the presence of IT. Web activist conflict occurs far from a battlefield, in a mediascape in which the United States rapidly lost the moral-emotional advantage of 12 September 2001.[35] In this environment, the Web accelerates and intensifies the transgression, reshaping, and transformation of the boundaries of identity and politics, especially since 'the instant we develop a new technology of communication – talking drums, papyrus scrolls, books, telegraph, radios,

televisions, computers, mobile phones – we at least partially reconstruct the self and its world, creating new opportunities for reflection,
perception, and social experience.'[36] It is in the context of a battle over
perceptions that the impact of IT on IR will be measured. The impact is
content.

Appendix 1: Methodology of *Insurgency Online*

Criteria for Website Analysis

Documents

- How many documents are on the website?
- What types of documents are on the website and how many are there of each? Types might include press releases (often called communiqués), analyses, theoretical texts, texts by other organizations or groups (e.g., news reports, statements by governments, statements by allied groups).
- What is the subject of the texts? The group itself? A situation? National or international issues? Social or political issues? Group the texts by issue areas that they address (e.g., 'on the situation in country x'). Specifying the issues that are referred to will indicate the concerns of each group.
- Does the website address a conflict?
- Does the group view the country's regime as legitimate? Why or why not?
- Does the group describe itself and its goals? If so, how (e.g., liberal-democratic, socialist-communist, Islamic, nationalist, etc.)?
- How many texts are there? What are their lengths? Are some types of texts longer than others? Are some types of texts more frequent than others? This will indicate what the group is using the website for.
- Dates of the texts: What is the frequency of texts in a given year? Does the frequency of texts relate to political developments at a national or international level?

- What is the response to non-domestic issues? – i.e., does the group have much to say about events outside its immediate physical location?
- Are the friends or enemies of the group mentioned? Are materials by friends or enemies linked into the site?

Functions

- Location of the site – is there a country server (e.g., '.ca,' '.se,' etc.).
- Is the site a stand-alone document or affiliated with some umbrella organizations? For example, http://www.hamas.org is stand-alone, but http://sunsite.unc.edu/freeburma/index.html is not. Hamas has a domain name, which might indicate how well organized, financed, and connected the group is.
- Trace the site to its Web server – e.g., http://www.yorku.ca/research/ionline/insuron.htm can be traced to York University and viewers can see if there are other similar resources linked to that site.
- Who is the webmaster and where is it located? Is it located in the same country? The same continent? Are they part of a larger organization?
- What functions are provided on the site? – mailing lists? newsgroups? chatrooms? video or audio documents? photo images (of whom or what, how many, why)? How many functions are provided?
- What other miscellaneous or distinct features are built into the site?

Criteria for Country Analysis

- Population – homogeneous group or many ethnicities/linguistic groups?
- Economy – resources, industries, sources of income, poverty vs. wealth, etc.
- Type of political system – military regime, liberal democracy, communist, Islamic state, dictatorship, etc.
- What is the particular structure of the regime (e.g., federal, unitary, parliamentary, presidential)?
- How do political structures allow or not allow for difference and identity (e.g., a multi-ethnic state might have federalism or special representation for minorities). What structures are in place?
- Status of media – are constitutional guarantees for freedom of

expression provided and respected by the powers that be? Is there a relatively open media system?

- Political context – type of political system (remember that everyone says they are democratic, but what does it mean to say that North Korea is democratic?).
- Political players – Does a legal opposition exist? Is it legal to oppose/criticize the regime? Who are the opposition groups?
- Issues – What are the political issues that underlie whatever conflict or situation is underway in this example?

Resources*

- Country Studies: Area Handbook Series (Library of Congress, Federal Research Division) http://lcweb2.10c.gov/frd/cs/cshome.html.
- Jane's Defence (especially the geopolitical section) – http://defence.janes.com/.
- Political Resources on the Net. This is an extremely useful resource for locating political materials on the Web: http://www.politicalresources.net/.
- Amnesty International – http://www.amnesty.org/.
- 'The Internet under Surveillance – Obstacles to the Free Flow of Information Online,' Reporters Sans Frontières – http://www.rsf.fr/article.php3?id_article=7280, 19 June 2003 (accessed 26 March 2005).
- Bureau of Democracy, Human Rights, Labor, U.S. State Department – http://www.state.gov/g/drl/.
- Human Rights Watch – http://www.hrw.org.
- UN High Commissioner for Human Rights (UNHCHR) – http://www.unhchr.ch/.
- UN High Commissioner for Refugees (UNHCR) – http://www.unhcr.ch/.
- U.S. Agency for International Development (USAID) http://www.info.usaid.gov/.

---

* The following websites were accessed in August 2000.

Appendix 2: Some of the Restrictions Imposed by the Taliban on Women in Afghanistan*

The following list offers only an abbreviated glimpse of the hellish lives Afghan women are forced to lead under the Taliban, and cannot begin to reflect the depth of female deprivations and sufferings. Taliban treat women worse than they treat animals. In fact, even as Taliban declare the keeping of caged birds and animals illegal, they imprison Afghan women within the four walls of their own houses. Women have no importance in Taliban eyes unless they are occupied producing children, satisfying male sexual needs or attending to the drudgery of daily housework. Jehadi fundamentalists such as Gulbaddin, Rabbani, Masood, Sayyaf, Khalili, Akbari, Mazari and their co-criminal Dostum have committed the most treacherous and filthy crimes against Afghan women. And as more areas come under Taliban control, even if the number of rapes and murders perpetrated against women falls, Taliban restrictions – comparable to those from the middle ages – will continue to kill the spirit of our people while depriving them of a humane existence. We consider Taliban more treacherous and ignorant than Jehadis. According to our people, 'Jehadis were killing us with guns and swords but Taliban are killing us with cotton.'

Taliban restrictions and mistreatment of women include the:

1. Complete ban on women's work outside the home, which also applies to female teachers, engineers and most professionals. Only a few female doctors and nurses are allowed to work in some hospitals in Kabul.

---

*Source: RAWA at http://songs.rawa.org/rawa/rules.htm (accessed 22 August 2001).

2. Complete ban on women's activity outside the home unless accompanied by a mahram (close male relative such as a father, brother or husband).
3. Ban on women dealing with male shopkeepers.
4. Ban on women being treated by male doctors.
5. Ban on women studying at schools, universities or any other educational institution. (Taliban have converted girls' schools into religious seminaries.)
6. Requirement that women wear a long veil (Burqa), which covers them from head to toe.
7. Whipping, beating and verbal abuse of women not clothed in accordance with Taliban rules, or of women unaccompanied by a mahram.
8. Whipping of women in public for having non-covered ankles.
9. Public stoning of women accused of having sex outside marriage. (A number of lovers are stoned to death under this rule).
10. Ban on the use of cosmetics. (Many women with painted nails have had fingers cut off).
11. Ban on women talking or shaking hands with non-mahram males.
12. Ban on women laughing loudly. (No stranger should hear a woman's voice).
13. Ban on women wearing high heel shoes, which would produce sound while walking. (A man must not hear a woman's footsteps.)
14. Ban on women riding in a taxi without a mahram.
15. Ban on women's presence in radio, television or public gatherings of any kind.
16. Ban on women playing sports or entering a sport center or club.
17. Ban on women riding bicycles or motorcycles, even with their mahrams.
18. Ban on women's wearing brightly colored clothes. In Taliban terms, these are 'sexually attracting colors.'
19. Ban on women gathering for festive occasions such as the Eids, or for any recreational purpose.
20. Ban on women washing clothes next to rivers or in a public place.
21. Modification of all place names including the word 'women.' For example, 'women's garden' has been renamed 'spring garden.'
22. Ban on women appearing on the balconies of their apartments or houses.
23. Compulsory painting of all windows, so women can not be seen from outside their homes.

24. Ban on male tailors taking women's measurements or sewing women's clothes.
25. Ban on female public baths.
26. Ban on males and females traveling on the same bus. Public buses have now been designated 'males only' (or 'females only').
27. Ban on flared (wide) pant-legs, even under a burqa.
28. Ban on the photographing or filming of women.
29. Ban on women's pictures printed in newspapers and books, or hung on the walls of houses and shops.

Apart from the above restrictions on women, the Taliban has:

- Banned listening to music, not only for women but men as well.
- Banned the watching of movies, television and videos, for everyone.
- Banned celebrating the traditional new year (Nowroz) on March 21. The Taliban has proclaimed the holiday un-Islamic.
- Disavowed Labor Day (May 1st), because it is deemed a 'communist' holiday.
- Ordered that all people with non-Islamic names change them to Islamic ones.
- Forced haircuts upon Afghan youth.
- Ordered that men wear Islamic clothes and a cap.
- Ordered that men not shave or trim their beards, which should grow long enough to protrude from a fist clasped at the point of the chin.
- Ordered that all people attend prayers in mosques five times daily.
- Banned the keeping of pigeons and playing with the birds, describing it as un-Islamic. The violators will be imprisoned and the birds shall be killed. The kite flying has also been stopped.
- Ordered all onlookers, while encouraging the sportsmen, to chant Allah-o-Akbar (God is great) and refrain from clapping.
- Ban on certain games including kite flying which is 'un-Islamic' according to Taliban.
- Anyone who carries objectionable literature will be executed.
- Anyone who converts from Islam to any other religion will be executed.
- All boy students must wear turbans. They say 'No turban, no education.'
- Non-Muslim minorities must wear a distinct badge or stitch a

yellow cloth onto their dress to be differentiated from the majority Muslim population. Just like what Nazis did with Jews.
- Banned the use of the Internet by both ordinary Afghans and foreigners.

And so on ...

Appendix 3: Translation of the *Du'a* of Sheikh Muhammed Al Mohaisany

Oh Allah, to whom belongs All Glory and Grandeur, Oh Allah The Omni-Potent, The Supreme, The Greatest, The Highest We ask you of Your Glory and Power, And victory for the Mujahideen in your cause, Oh Allah remain beside them, and with them, Give them triumph; strengthen them. Oh Allah unite their vision, Focus the aim of their weaponry, And consolidate their word.

And O Allah, fix their hearts, O Allah handle, take care of their enemies, O Allah dissipate their congregation, And shatter their integrity, And weaken their strengths, And throw the fear in their hearts, O Allah, our fates are in your hands, And our affairs all return to You, And our conditions are not obscured from Your knowledge, To You do we raise our misery, And our sorrow, And our complaint, To You, and You alone, do we complain the injustice of the oppressors, And the cruelty of the 'faajereen" (literally: perpetrators of debauchery), And the wrath of the betraying criminals,

To You, O Allah, do we complain the injustice of the spiteful Christians, O Allah, the night (the dark reign) of the oppressors has indeed lengthened, O Allah, the night of the oppressors has indeed lengthened, ... O Allah, the night of the oppressors has indeed lengthened, O Allah, the night of the oppressors has indeed lengthened The animosity of the atheists has extended deep

And ... the heads of the criminals Oh Allah, Oh Allah, Send upon them a hand from the truth ... To raise with it our humiliation, And to return to us our dignity, And to destroy our enemy with it, Oh Allah, Oh

Allah, take care of the sources of injustice and oppression, Oh Allah, take care of the sources of injustice and oppression, Oh Allah, direct your forces against America, The center of Kufr and Fasaad Oh Allah, direct your forces against America, The center of Kufr and Fasaad Oh Allah, of them our are All-Aware, They spread fasaad in Your lands,

And they killed Your slaves, And they insulted Your religion Oh Allah, of them our are All-Aware, And over them All-Powerful, Oh Allah, direct your forces against them, Oh Allah, direct your forces against them, O Allah send upon them the Storms of 'Aad, And the Cry of the Thamoud, And the Typhoon of the people of Noah, O Allah send upon them that which descends from the skies, And of that which exudes from the lands, O Allah disintegrate their country, O Allah make them into divided countries and scattered parties, O Allah, Ever-Living and Omni-Potent, Make contain them within a fist's grip of Your slaves (i.e. under their control), Make contain them within a fist's grip of Your slaves, O Allah, make hurricanes a constant for them,

O Allah, make hurricanes a constant for them,

O Allah release our captured brothers, O Allah release them, O Allah, strengthen them, O Allah make them steady on their faith, O Allah make possible a means for them, O Allah handle those who torture them,

O Allah handle those who torture them, O Allah handle those who torture them O Allah eradicate them with Your power and Omni-Potence, O Allah, make their plots against us a cause for their destruction, And their slyness, slyness against them, O Lord of the worlds, O Ever-Living, O Omni-Potent, O Most-Mighty and Most-Gracious, Hearer of all prayer, Ever so close, accepting to all prayers, We all pray to You, full aware of Your promise, And of Your acceptance, For You have said, and Your speech is the truth: 'Pray to Me, for I accept your prayers,' O Allah accept our prayers for us, O Allah accept our prayers for us, O Allah, Everlasting, All-Powerful, The Omni-Potent over all that is in the heavens and the earth,

We ask you to save Al-Aqsa from the cruelty of the Jews, O Allah save Al-Aqsa from the cruelty of the Jews, O Allah free Al-Aqsa from every black-hearted Kafir, O Allah lay our eyes rest on a liberated Aqsa, and on the defeat of the spiteful Jews, O Most-Mighty and Most-Gracious, All creatures are unto You humiliated, meek, ... O Allah, our creator from a single soul, O Allah, Highest in status O Allah, Greatest in Strength We ask You glory for Islam and Muslims, O Allah, Ever-Living, Ever so Powerful, O Allah guard the hard working scholars, O Allah guard the hard working scholars,

And make steadfast those sincere in inviting to Your path, And raise the positions of those who order righteousness, and who forbid evil, And bestow the same mercy upon those Muslims who enjoin them, O Allah, Ever-Living, Ever so Powerful, O Most-Mighty and Most-Gracious, O Allah, he who devoted himself to hurt them, Talking to defame their honor, And tracking their refuge, And for whom You have willed no guidance, O Allah, make misery his destiny, O Allah, make misery his destiny, And disaster in his path, O Allah, convert his health to disease, And his strength into sickness, And his wealth, into poverty, And his power into weakness.

Notes

1 Introduction: Insurgency Online and Conflict in the Global-scape

1 Graeme Smith, 'Pentagon Downed Web Site, Al-Jazeera Editor Says,' *Globe and Mail*, 29 March 2003, A6. See as well 'Hackers Cripple al-Jazeera Sites,' BBC News, http://news.bbc.co.uk/2/hi/technology/2893993.stm (accessed 30 March 2003).

2 Arjun Appadurai, 'Disjuncture and Difference in the Cultural Economy,' in *Planet TV: A Global Television Reader*, ed. Lisa Parks and Shanthi Kumar (New York: New York University Press, 2003), 42–3.

3 See, for example, Stephen Duncombe, ed., *Cultural Resistance Reader* (London: Verso, 2002).

4 Ronald Deibert, 'Altered Worlds: Social Forces in the Hypermedia Environment,' in *Digital Democracy*, ed. Cynthia Alexander (Toronto: Oxford University Press, 1998).

5 All statistics cited here are from the website of the International Telecommunication Union (ITU), http://www.itu.int/ITU-D/ict/statistics/ (accessed 13 August 2003).

6 See ibid.

7 See Brigitte Lebens Nacos, *Mass-Mediated Terrorism: The Central Role of the Media in Terrorism and Counterterrorism* (Lanham, MD: Rowman and Littlefield, 2002).

8 For a more nuanced view, see Piers Robinson, *The CNN Effect: The Myth of News, Foreign Policy and Intervention* (London: Routledge, 2002).

9 See Michael Y. Dartnell, *Action Directe: Ultra-Left Terrorism in France, 1979–1987* (London: Frank Cass, 1995).

10 See Danny Schechter, *Media Wars: News at a Time of Terror* (Lanham, MD: Rowman and Littlefield, 2003), and Joseba Zulaika and William A.

Douglass, *Terror and Taboo: The Follies, Fables, and Faces of Terrorism* (New York: Routledge, 1996).

11 See Appendix 1: Methodology of *Insurgency Online*.

12 'U.S. Government Electronic Commerce Policy,' http://www.ecommerce. gov/ (accessed 28 March 2000).

13 See the New York Stock Exchange website at http://www.nyse.com/ as an example of how information is influencing international economics (accessed 28 March 2000).

14 'E-Stats,' U.S. Department of Commerce, U.S. Census Bureau, Economic and Statistics Adminisration, 18 March 2002, www.census.gov/estats (accessed 5 November 2002).

15 'Measuring Electronic Business: Definitions, Underlying Concepts and Measurement Plans,' Thomas L. Mesenbourg, Assistant Director for Economic Programs, Bureau of the Census, 13 October 1999, http:// www.ecommerce.gov/ecomnews/e-def.html (accessed 16 November 1999): 'While the burgeoning use of electronic devices in our economy is widely acknowledged and discussed, it remains largely undefined and unrecognized in official economic statistics. The terms Internet, electronic commerce, electronic business, and cybertrade are used often. However, they are used interchangeably and with no common understanding of their scope or relationships. Establishing terms that clearly and consistently describe our growing and dynamic networked economy is a critical first step toward developing useful statistics about it.'

16 The problem was acknowledged by the U.S. Department of Commerce under the Clinton administration, which created a website focused on the subject: 'Understanding the Digital Economy,' http://www. digitaleconomy.gov/ (accessed 28 March 2000).

17 For example, I now routinely read the *The Guardian* at http://www. guardian.co.uk, which a decade ago was impractical and unattractive due to availability and cost.

18 Mark Alleyne, *International Power and International Communication* (London: Macmillan, 1995).

19 See especially Dan Schiller, 'How to Think about Information,' in *The Political Economy of Information*, ed. Vincent Mosco and Janet Wasco (Madison: University of Wisconsin Press, 1988), 27–43.

20 Robert O. Keohane and Joseph S. Nye, Jr, 'Power and Interdependence in the Information Age,' *Foreign Affairs* 77, no. 5 (September-October 1998): 81–92.

21 Catherine Alexander, 'National Security in a Wired World,' in *Digital*

Democracy: Policy and Politics in the Wired World, ed. Cynthia Alexander and Leslie Pal (Toronto: Oxford University Press, 1998), 46–62.

22 See Howard Frederick, *Global Communication and International Relations* (Orlando, FL: Harcourt Brace, 1993), and Nick Stevenson, *The Transformation of the Media: Globalisation, Morality and Ethics* (London: Longman, 1999).

23 See Jay David Bolter and Richard Grusin, *Remediation: Understanding New Media* (Cambridge, MA: MIT Press, 1999).

24 Other recent works in the area of IT and IR include: Edward Comor, *Communication, Commerce and Power: The Political Economy of America and the Direct Broadcast Satellite, 1960–2000* (London and New York: Macmillan and St Martin's Press, 1998), *The Global Political Economy of Communication: Hegemony, Telecommunication and the Information* (New York: Palgrave Macmillan, 1996), 'The Role of Communication in Global Civil Society: Forces, Processes, Prospects,' *International Studies Quarterly* 45, no. 3 (September 2001): 309–23, 'New Technologies and Consumption: Contradictions in the Emerging World Order' in *Information Technologies and Global Politics: The Changing Scope of Power and Governance*, ed. James Rosenau and J.P. Singh (Albany: SUNY Press, 2002), 169–85, 'Governance and the Nation-State in a Knowledge-Based Political Economy,' in *Approaches to Global Governance Theory*, ed. Timothy J. Sinclair and Martin Hewson (Albany: SUNY Press, 1999), 117–34; Peter Cowhey, 'The International Telecommunications Regime: The Political Roots of Regimes for High Technology,' *International Organization* 44, no. 2 (1990): 169–200; Ronald Deibert, *Parchment, Printing, and Hypermedia* (New York: Columbia University Press, 1997); James Der Derian, *Virtuous War: Mapping the Military-Industrial-Media-Entertainment Network* (Boulder, CO: Westview Press, 2001), 'Global Events, National Security, and Virtual Theory,' *Millennium Journal of International Studies* 30, no. 3 (2001): 669–90; Karl Deutsch, *The Nerves of Government: Models of Political Communication and Control* (New York: Free Press, 1969); William Drake, *The New Information Infrastructure: Strategies for US Policy* (Washington, DC: The Brookings Institution, 1995); Cees Hamelink, *The Politics of World Communication* (Thousand Oaks, CA: Sage, 1994); Stephen Krasner, 'Global Communications and National Power: Life on the Pareto Frontier,' *World Politics* 43, no. 3 (1991): 336–66; Robert Latham, *Bombs and Bandwidth: The Emerging Relationship between Information Technology and Security* (New York: The New Press, 2003); Ithiel de Sola Pool, *Technologies without Boundaries: On Telecommunications in a Global Age* (Cambridge: Harvard University Press, 1990).

25 'Strategic Assessment: The Internet,' Charles Swett, Office of the Assistant

Secretary of Defense for Special Operations and Low-Intensity Conflict, July 1995, www.isoc.org/inet96/proceedings/e1/e1_2.htm: 'The Internet has been increasingly involved in politics and international conflict. Local, state and national governments are establishing a presence on the Internet, both for disseminating information to the public and for receiving feedback from the public. Candidates for elective office are conducting debates over the Internet. Organizers of domestic and international political movements are using the Internet. It has played a key role in Desert Storm, the Tiananmen Square massacre, the attempted coup in Russia, the conflict in the former Yugoslavia, and in the challenge to authoritarian controls in Iran, China, and other oppressive states. The Internet is playing an increasingly significant role in international security, a role that is potentially important to DoD.'

26 See electronicintifada.org, electroniciraq.org, and aboutbaghdad.com.
27 Military manifestations of sovereignty and by implication sovereignty itself have long been the subject of satire, some of which is now available online. See 'puppet soldiers' at http://cain.ulst.ac.uk/bogsideartists/bsprotest2.htm (accessed 29 March 2000).
28 Monroe Price, *Media and Sovereignty: The Global Information Revolution and Its Challenge to State Power* (Cambridge, MA: MIT Press, 2002), 26.
29 Saskia Sassen, *Globalization and Its Discontents: Essays on the New Mobility of People and Money* (New York: New Press, 1998), 99.
30 James Rosenau, *Distant Proximities: Dynamics Beyond Globalization* (Princeton: Princeton University Press, 2003), 4.
31 Appadurai, 'Disjuncture and Difference,' 41.
32 Fen Osler Hampson, *Madness in the Multitude: Human Security and World Disorder* (Toronto: Oxford University Press, 2002), 174.
33 See Benedict Anderson, *Imagined Communities* (London: Verso, 1983).
34 See, for example, http://www.akaKURDISTAN.com (accessed 28 March 2000).
35 See, for example, IRAradio.com (Irish Republican Activist Radio), http://www.iraradio.com/ (accessed 28 March 2000).
36 Dorothy Denning and Peter Denning, *Internet Besieged: Countering Cyberspace Scofflaws* (New York: Addison-Wesley, 1998), 51.
37 Keohane and Nye, Jr, point this out in 'Power and Interdependence in the Information Age.'
38 Shanthi Kalathil and Taylor Boas, *Open Networks, Closed Regimes: The Impact of the Internet on Authoritarian Rule* (Washington, DC: Carnegie Endowment for International Peace, 2003), 132–3.
39 An example of this is the website of the Chechen Republic site at http://

www.amina.com/ (accessed 8 August 2003). While the Chechen struggle for independence is not new, the appeal to a global public has sparked widespread awareness of the situation and the various human rights violations of the Russian authorities in the breakaway region.

40 John Arquilla and David Ronfeldt, 'Cyberwar Is Coming!' *Comparative Strategy* 12, no. 2 (Spring 1993): 141.

41 Ibid., 144.

42 This was seen in the emergence of the Zapatistas in Chiapas, Mexico, in the 1990s. See their website 'YA BASTA! (Zapatista National Liberation Army)' at http://www.ezln.org/ (accessed 29 March 2000).

43 Arquilla and Ronfeldt, *The Zapatista Social Netwar in Mexico* (Santa Monica, CA: RAND, 1998), 114.

44 Arquilla and Ronfeldt, *The Advent of Netwar* (Santa Monica, CA: RAND, 1996), 13.

45 The Spanish-language MRTA website is online at http://www.nadir.org/nadir/initiativ/mrta/ (accessed 15 August 2003).

46 See USA Patriot Act HR 3162 (Uniting and Strengthening America by Providing Appropriate Tools Required to Intercept and Obstruct Terrorism (USA PATRIOT ACT) Act of 2001), http://www.hqda.army.mil/rio/links/USA%20PATRIOT%20Act%20HR%20316 2.htm (accessed 25 July 2003).

47 See Michael Y. Dartnell, 'Multimedia as Transgressive Practice: The D'ua of Sheikh Muhammed Al Mohaisany,' in *Governance and Global (Dis)orders: Trends Transformations and Impasses*, ed. Alison Howell (Toronto: York Centre for International and Security Studies, 2004), 289–312.

48 This can be seen in Northern Ireland political wall murals that are reproduced on the Web. See 'King Billy Mural' at http://cain.ulst.ac.uk/bibdbs/murals/slide1.htm#1 (accessed 29 March 2000).

49 Chris Hables Gray, *Postmodern War: The New Politics of Conflict* (New York: Guilford Press, 1997), 22.

50 Mark S. Bonchek, 'Grassroots in Cyberspace: Using Computer Networks to Facilitate Political Participation,' MIT Artificial Intelligence Laboratory Working Paper 95–2.2, presented at 53rd Annual Meeting of the Midwest Political Science Association Chicago, IL, http://www.ai.mit.edu/people/msb/pubs/grassroots.html (accessed 6 April 1995).

51 BBC News, 'India's Digital Divide,' 25 May 2003, http://news.bbc.co.uk/2/hi/programmes/from_our_own_correspondent/ 2932758.stm (accessed 9 August 2003).

52 The differences in Internet access are obvious at a global level. According to the ITU, there are large discrepancies between different global regions.

While there are only 1.23 personal computers for every 100 inhabitants in Africa, there are 27.49 in North and South America, 3.95 in Asia, 20.01 in Europe, and 38.94 in Oceania. As one study notes, 'the Internet does not touch the lives of most inhabitants of the planet. In spite of its growth over the past decade, the Internet is used by less than 10% of the world's population, making it less widespread than the telephone or television or radio. Moreover, its growth is non-uniform. In the vast majority of countries, fewer than one percent of the population uses the Internet.' Peter Wolcott, 'Introducing the Global Diffusion of the Internet Series,' *Communications of the Association for Information Systems* 11 (2003): 555–9, 556.

53 Anthony H. Richmond, *Global Apartheid: Refugees, Racism, and the New World Order* (Toronto: Oxford University Press, 1994), 31.

54 Manuel Castells, *The Rise of the Network Society* (Malden, MA: Blackwell, 1996), 364.

55 W. Rash, *Politics on the Nets: Wiring the Political Process* (New York: W.H. Freeman, 1997), 21.

56 Sending large quantities of unsolicited email is called 'spamming.'

57 When email messages emanating from a computer in the Balkans started to overload NATO's website during the 1999 Kosovo campaign, the attacks were probably neutralized by just such a procedure. The danger then would be on the other foot: NATO would be able to trace the source of 'ping bombardment.'

58 Changing a message and resending it while pretending to be part of the organization that created it is called 'spoofing.'

59 Information on white supremacist and hate groups and their activities on the Web can be found at the websites of the Southern Poverty Law Center, http://www.splcenter.org/; the Simon Weisenthal Center http://www.wiesenthal.com/; Tolerance.org, http://www.tolerance.org/; and the Anti-Defamation League, http://www.adl.org/. A list of hate-based groups active in the United States is available at http://www.tolerance.org/maps/hate/index.html (accessed 7 August 2003).

60 Joel R. Reidenberg, 'Governing Networks and Rule-Making in Cyberspace,' in *Borders in Cyberspace: Information Policy and the Global Information Infrastructure*, ed. Brian Kahin and Charles Nesson (Cambridge, MA: MIT Press, 1997), 86.

61 While this particular blackout did not affect the workings of the Internet, it effectively knocked millions of users offline by bringing down computer systems and telephones.

62 Richmond, *Global Apartheid*, 200.

63 Castells, *Rise of the Network Society,* 99.
64 Peter J. Anderson, *The Global Politics of Power, Justice and Death* (London: Routledge, 1996), 76.
65 U.S. parallels include Franklin Delano Roosevelt's successful use of radio and the famous televised debates between John Fitzgerald Kennedy and Richard Nixon.
66 'Les premiers mois de la Révolution voient l'extraordinaire explosion des publications périodiques: 42 titres entre mai and juillet 1789, plus de 250 titres pour le second semestre de 1789. Cette explosion a été préparée, depuis la réunion de l'Assemblée des notables, par la multiplication des libelles, dont certain paraissent en plusieurs fascicules, à intervalles réguliers, annonçant ainsi les périodiques. Elle est facilitée par les conditions de fabrication: le matériel nécessaire à la publication d'un journal ne coûte pas cher; une seule et même personne peut diriger, rédiger, imprimer, vendre un journal. La feuille révolutionnaire, écrite sur du méchant papier, bourrée de fautes d'impression, est une fabrication aussi haletante que l'actualité qu'elle commente, et presque toujours liée à la personnalité d'un unique rédacteur, qui ne signe pas des articles, mais fait tout le journal.' Mona Ozouf, 'Esprit public,' in *Dictionnaire critique de la Révolution française,* ed. François Furet and Mona Ozouf (Paris: Flammarion, 1988), 714; my translation.
67 Dale K. Van Kley, 'New Wine in Old Wineskins: Continuity and Rupture in the Pamphlet Debate of the French Revolution,' *French Historical Studies* 17, no. 2. (1991): 460.
68 Mirabeau wrote that newspapers 'establish communications that cannot fail to produce harmony of sentiments, of opinions, of plans, and of action that constitutes the real public force, the true safeguard of the constitution.' Comte de Mirabeau, *Les États-généraux,* Le Jay, publisher.
69 Jacques-Pierre Brissot, 1789, cited in Ozouf, 'Esprit public,' 10.
70 On 5 July 1788, Louis XVI invited 'all learned and educated persons in the kingdom' to write 'papers' and 'information' about the state of current affairs. This led to avalanche of publications, including the first independent newspaper such as Brissot's *Le Patriote français* and Mirabeau's *Les États-généraux*. About 500 publications appeared between 1789 and 1792. On 26 August 1789, Article 11 of the Declaration of the Rights of Man and of the Citizens stated that 'free communication of ideas and opinions is one of the most precious of the rights of man. Every citizen may, accordingly, speak, write, and print with freedom, but shall be responsible for such abuses of this freedom as shall be defined by law.' Human and Constitutional Rights Resource Page, www.hrcr.org/docs/frenchdec.html.

71 Other examples of the effects of the eighteenth-century print revolution were British Wilkite agitation to attract popular support for a campaign against oligarchic rule in the 1760s and the United Irishmen campaign of the 1790s. See John Brewer, *Party Ideology and Popular Politics at the Accession of George III* (Cambridge: Cambridge University Press, 1981), and Nancy J. Curtin, *The United Irishmen: Popular Politics in Ulster and Dublin 1791–1798* (Oxford: Clarendon Press, 1998).

72 See Bolter and Grushin, *Remediation*.

73 See Kenneth Plummer, *Telling Sexual Stories: Power, Change and Social Worlds* (London: Routledge, 1995).

74 Mark Poster, 'CyberDemocracy: Internet and the Public Sphere,' http://www.hnet.uci.edu/mposter/writings/democ.html (1995).

75 Kiku Adatto, *Picture Perfect: The Art and Artifice of Public Image Making* (New York: Basic Books, 1993), 12.

76 Ibid., 18.

77 One countervailing tendency is the rising power of the Chinese state. Other trends include the Bush administration's invasion of Iraq in 2003, the Indian government's May 1998 nuclear tests, and nuclear tests by France in 1995, all of which show that globalization might not lead states to relinquish sovereignty.

78 Bertolt Brecht, for example, recognized the value of radio communication in politics in 1932. See chapter 5 below.

79 'Referring to a U.S. Chamber of Commerce event yesterday demonstrating the use of the Internet as a lobbying tool for communicating with (lobbying) government officials, James Thurber, director of American University's Center for Congressional Studies, says: "This is an example of the future ... The linkage between a direct lobbying effort and the Internet is going to improve the capacity of these large organizations to pressure individual members of Congress to do what they want them to do. With these sites, they can just click an icon, and they have programs that will automatically send a letter to the right members of Congress."' *Washington Post*, 18 May 1998.

80 Castells, *Rise of the Network Society*, 476.

81 David V.J. Bell, 'Global Communications, Culture, and Values: Implications for Global Security,' in *Building a New Global Order: Emerging Trends in International Security*, ed. David Dewitt, David Haglund, and John Kirton (Toronto: Oxford University Press, 1993, 172.

82 Susan Sontag, *On Photography* (New York: Picador, 2001), 149.

83 Joshua Goldstein, *International Relations*, 4th ed. (New York: Longman, 2001), 464–5.

84 Kalathil and Boas, *Open Networks*, 13.

85 Jean Baudrillard, 'Global Debt and Parallel Universe' (1997), http://www.ctheory.text_file.asp?pick=164 (accessed 25 March 2005).

86 James E. Katz and Ronald E. Rice, *Social Consequences of Internet Use: Access, Involvement and Interaction* (Cambridge, MA: MIT Press, 2002), 352.

87 The contents of the Sinn Fein (SF) mailing list are a revealing case in point. A main point of discussion over the 10 April 1998 Belfast agreement (assuming participants represent any significant body of opinion within SF) was the possibility that it might undermine the sovereignty of a future united Ireland.

88 The July 1997 handover of Hong Kong from Great Britain to the People's Republic of China highlights the continued importance of the state. While incorporating Hong Kong into China was in some sense a triumph of a global value of decolonization, the Chinese government portrayed the event as a national victory. The prevailing image is one of increasing complexity. The example of Hong Kong shows the complexity of global politics and the place of the state in contemporary power networks. The Chinese takeover in Hong Kong highlighted the importance of a very powerful state that has been built within the past fifty years. On the other hand, the handover was also a victory for decolonization, a transnational movement that has been prominent since at least 1919. In this light, it is hard to definitively assess the impact of Web activism on sovereignty beyond stating that it is one component in a complex global system.

89 Castells, *Rise of the Network Society*, 3.

90 Poster, 'CyberDemocracy: Internet and the Public Sphere.'

2 Insurgency Online as Networking: IRSM Web Activism

1 An early version of this chapter was presented at the International Political Science Association Quebec Meeting in Quebec City in August 2000. I would like to thank Geoff Kennedy for his excellent research on the IRSM website. A later version was published as 'The Electronic Starry Plough: The Enationalism of the Irish Republican Socialist Movement (IRSM)' in *First Monday* (Chicago: University of Illinois at Chicago), vol. 6, no. 12 (December 2001), http:firstmonday.org/issues/issue6_12/dartnell/index.html.

2 Benedict Anderson, *Imagined Communities: Reflections on the Origin and Spread of Nationalism* (London: Verso, 1983), 7.

3 Tom Nairn, *Face of Nationalism: Janus Revisited* (London: Verso, 1997), 17.

4 David Miller, *On Nationality* (Oxford: Clarendon Press, 1995), 185.

5 Ibid., 15.

6 This was notably the case of Irish nationalism, which developed overseas communities in the United States and Australia. In 2004, Ukrainian Canadians played a prominent role in backing the Orange Revolution.

7 See Michel Foucault, 'Security, Territory, and Population,' in *Michel Foucault: Ethics, Subjectivity and Truth*, ed. Paul Rabinow (New York: New Press, 1994), 69.

8 Robert Burnett and P. David Marshall, *Web Theory* (London: Routledge, 2003), 166–7.

9 See David Held, Anthony McGrew, David Goldblatt, and Jonathan Perraton, *Global Transformations: Politics, Economics and Culture* (Stanford, CA: Stanford University Press, 1999).

10 David Held, 'Violence, Law and Justice in a Global Age,' *After Sept. 11 Archive*, Social Science Research Council, 2001, http://www.ssrc.org/sept11/essays/held.htm (accessed 16 August 2003).

11 The 'Starry Plough' refers to the constellation Ursa Major, the Great Bear or Plough of the heavens.

12 Anthony Cohen, *The Symbolic Construction of Community* (New York: Routledge, 1985), 14.

13 In this discussion, 'Republic' refers to the Republic of Ireland. Non-capitalized instances of 'republic' or a 'workers' republic' do not refer to the Republic of Ireland.

14 Based on a 1991 census. Projections on the Catholic-Protestant mix are controversial in Northern Ireland. Current birth rates suggest a Catholic majority in the North around 2010.

15 Manufacturing output was up and GDP grew by 11.6 per cent in 1990–5. Unemployment is 11.2 per cent, but an improvement from highs of 17 per cent in the 1980s. Once a shipbuilding and linen producing centre, Northern Ireland was a net importer of most goods and exported little by the 1970s, when the welfare state and a massive security presence were the region's main economic stand-bys. Figures from 'Northern Ireland Economic Overview,' October 1997, Northern Ireland Office, http://www.nio.gov.uk/970919.htm (accessed 25 July 2000).

16 See Henry Patterson, 'Northern Ireland Economy,' in *Northern Ireland Politics*, ed. A. Aughey and D. Morrow (London: Longman, 1996), 127.

17 Joseph Ruane and Jennifer Todd, *The Dynamics of Conflict in Northern Ireland* (Cambridge: Cambridge University Press, 1996), 170.

18 Political parties include: Ulster Unionist Party (UUP), Democratic Unionist Party (DUP), Social Democratic and Labour Party (SDLP), Sinn Fein (SF), Alliance Party (AP), Northern Ireland Labour Party (NILP), Irish Republican Socialist Party (IRSP), Worker's Party (WP), Progressive Unionist Party (PUP), UK Unionist Party (UKUP), and Ulster Democratic Party (UDP).

19 Paul Arthur, 'Political Parties: Elections and Strategies,' in *Northern Ireland Politics*, ed. Aughey and Morrow, 18.

20 B. O'Leary and J. McGarry, *The Politics of Antagonism: Understanding Northern Ireland* (London: Athlone Press, 1993), 18.

21 L. Bairner, 'The Media,' in *Northern Ireland Politics*, ed. Aughey and Morrow, 19.

22 See *War and Words: The Northern Ireland Media Reader*, ed. Bill Rolston and David Miller (Belfast: Beyond the Pale Publications, 1996).

23 Figures from a report by P.J. O'Connell, R.O. O'Donnell, and V. Gash, *Astonishing Success: Economic Growth and the Labour Market in Ireland*, Dublin: The Economic and Social Research Institute, http://www.esri.ie/1999_BK_MN _SUM.HTM#AstonishingSuccess (accessed 25 July 2000). Figures for 2003 from the Central Statistics Office, Republic of Ireland, www.cso.ie (accessed 19 November 2005).

24 'Quarterly Economic Commentary,' Economic and Social Research Institute (ESRI), June 2000, http://www.esri.ie/QEC0600.HTM, (accessed 25 July 2000).

25 Economic and Social Research Institute, http://www.esri.ie/content.cfm?t=Irish%20Economy&mld=4 (accessed 22 December 2004).

26 See http://www.irsm.org/irsm.html.

27 'What Is National Liberation?' http://www.irsm.org/general/history/whatisnatlib.htm (accessed 25 July 2000).

28 IRSM, 'Why the IRSP?' http://www.irsm.org/general/history/whyirps.htm (accessed 25 July 2000).

29 Connolly's Marxism and activism are used by the IRSM as a symbol of a secular, union-based, and socialist Irish politics. Connolly was executed for his leading role in the 1916 Rebellion. He is often employed as a symbol of working-class militancy in Ireland and Northern Ireland. The IRSM's 'Connollyite Marxism' as well as the legacy of 'Green Marxism' in Ireland are found in 'What Is Irish Republican Socialism?' at http://www.irsm.org/general/history/whatis.htm and G Schuller's 'James Connolly and Irish Freedom' at http:www.irsm.org/general/history/jc&irishfreedom.htm (accessed 8 August 2003).

30 'Irish Republican Socialist Movement – 20 years of Struggle,' http:www.irsm.org/general/history/irsm20yr.htm (accessed 25 July 2000).

31 The IRSM position on the 'false consciousness' of the Northern Ireland working class are found in 'Loyalism,' http:www.irsm.org/general/history/loyalism.htm, and 'The Broad Front,' http:www.irsm.org/general/history/broadfront.htm (accessed 25 July 2000).

32 IRSM analysis of the relationship between the Troubles, British imperial-

ism, and capitalism is found in 'Capitalism,' http:www.irsm.org/general/ history/capitalism.htm (accessed 25 July 2000).

33 IRSM, 'The Road to Revolution,' http://www.irsm.org/general/history/ rtrinireland.htm (accessed 25 July 2000).

34 Ibid.

35 Ibid.

36 IRSM, 'The Broad Front.'

37 IRSM, 'Women in Ireland,' http:www.irsm.org/general/history/ women.htm (accessed 25 July 2000).

38 Ibid.

39 See http:www.irsm.org/statements/irsp.html and http:www.irsm.org/ statements/inla.html.

40 African Liberation Day, Kurdish New Year, and International Women's Day.

41 Solidarity for Kurdish Struggles and liberation in South Africa.

42 Actions condemned include: arrest of anti-NATO protesters in Italy, imprisonment of Mumia Abu-Jamal, the Kosovo war, Kurdish leader Abdullah Ocalan's 'abduction' and death sentence, arrest of Puerto Rican activists, Third World debt, the Pope's comments on Pinochet, sanction against Libya, U.S. extradition of IRSP POWs, and British 'acts of international terrorism.' The organization also calls for dissolution of NATO.

43 Scottish Republicans (The Scottish Republican Socialist Party), Cyrmu Goch-Welsh Socialists, and the Scottish Republican Forum.

44 Links are provided to the following websites: Free Éireann, Irish Republican Writer's Group, Scottish Republican Socialist Party, Scottish Socialist Alliance, 1913 Commemoration Committee, Militant Labour Home Page – Ireland, South African Communist Party, EZLN-Zapatistas, Celtic League, Committee on the Administration of Justice, FARC Home Page, Partido Democratico Popular Revolucionario y del Ejercito Popular Revolucionario (PDPR-EPR Mexico), Kurdistan Information Centre, Y Faner Goch (Cymru Goch's Periodical), Dublin Abortion Rights Alliance, Troops Out! Movement, Friends of Irish Freedom, Revolutionary People's Liberation Party Front (DHKP/C), Kurdish Worker's Party (PKK), Euskal Herria Journal (Basque), MRTA Solidarity Page, Romani Page, Sami People Page, and Widerstand Information and Analysis (German anti-fascist, anti-racist, socialist group).

45 These include: Radical History on the Web, Irish History on the Web, Larkspirit Online Bookshop, Ireland's Patriots, Venceremos Page, The Marxism/Leninism Project, List of Marxist Websites, Irish Struggles, The Irish Citizen Army: Labour clenches its fist, Ireland and British Imperialism, The

Marx/Engels archive, Mark's Solidarity Page (Irish-Mexican), and Spunk Press Home Page.

46 These documents are: 'Teach na Failte,' 'Seven Stars,' 'Communist Manifesto,' 'Articles by Jenny Marx on the Irish Question,' and 'If an Agent Knocks: Federal Investigators and Your Rights.'

47 James Rosenau, 'Information Technologies and the Skills, Networks, and Structures That Sustain World Affairs,' in *Information Technologies and Global Politics: The Changing Scope of Power and Governance* (Albany: SUNY Press, 2002), 284.

48 Manuel Castells, *The Rise of the Network Society* (Malden, MA: Blackwell, 1996), 476.

49 Burnett and Marshall, *Web Theory*, 2

50 David V.J. Bell, 'Global Communications, Culture, and Values: Implications for Global Security,' in *Building a New Global Order: Emerging Trends in International Security*, ed. David Dewitt, David Haglund, and John Kirton (Toronto: Oxford University Press, 1993), 172.

51 See Dan Schiller, 'How to Think about Information,' in *The Political Economy of Information*, ed. Vincent Mosco and Janet Wasco (Madison: University of Wisconsin Press, 1988), 27–43.

52 Castells, *The Rise of the Network Society*, 3.

3 Insurgency Online as Global Witnessing: The Web Activism of RAWA

1 Preliminary research on the RAWA website was conducted in July and August 2000. The remainder of the research was completed in August and September 2001. Additional materials that have since been added to the site are not examined here. References to the RAWA site are to the August 2001 version, which is used as a 'snapshot' of Web activism in the same way as the IRSM and MRTA websites. I would like to thank Sofie Tzoutzi for her excellent research on the RAWA website and Marnie Lucas-Zerbe for assembling a bibliography on RAWA and Afghan politics.

2 Aristotle, *The Politics*, ed Ernest Barker (London: Oxford University Press, 1974), bk I, chap. 1, ¶2, p. 8.

3 Ibid., bk I, chap. 6, ¶8, p. 16.

4 Jonathan Ned Katz, *The Invention of Heterosexuality* (New York: Plume, 1995), 31. Krafft-Ebing is a seminal figure for modern gender studies insofar as he defined two antithetical 'sexes.' This polarization is both instructive for political science and accurate in gender and sexuality studies because it is the basis for later writings by Foucault, Weeks, Butler, and others. In this sense, referring to Krafft-Ebing is similar to

citing Marx instead of Lenin, Luxemburg, Gramsci, and other followers of Marx.

5 A good source discussion regarding how theories of biological-sex difference were implemented as public policy in Canada is Gary Kinsman's *The Regulation of Desire* (Montreal: Black Rose Books, 1996). This process has also been researched by Foucault, Weeks, and others.

6 As an illustration, see Mariah Burton Nelson, *The Stronger Women Get, the More Men Love Football: Sexism and the American Culture of Sports* (New York: Harcourt Brace, 1994).

7 Among other things, '200,000 Koreans ... were forced to be sex slaves for the Japanese Imperial Army.' Sylvia Yu, 'Who Cares for Sex Slaves?' *Globe and Mail*, 17 August 2001, A11.

8 *Women, Peace and Security: Study Submitted by the Secretary-General Pursuant to Security Council Resolution 1325 (2000)* (UN, 2002), 3.

9 Jan Jindy Pettman, *Worlding Women: A Feminist International Politics* (London: Routledge, 1996), 160.

10 Ibid., 186–7.

11 Geraldine Heng, '"A Great Way to Fly": Nationalism, the State, and the Varieties of Third-World Feminism,' in *Feminist Genealogies, Colonial Legacies, Democratic Futures*, ed. M. Jacqui Alexander and Chandra Talpade Mohanty (New York: Routledge, 1997), 30.

12 Sally Armstrong, *Veiled Threat: The Hidden Power of the Women of Afghanistan* (Toronto: Viking, 2002), 7.

13 Mark MacKinnon, 'Progress Lagging for Afghan Women,' *Globe and Mail*, 6 August 2003, A10.

14 A tendency to view foreign conflicts as 'exotic' and unresolvable also infected the Bush administration's response to the Liberian conflict in 2003. It is possible that more sinister motives underlie the administration's failure to respond more rapidly to the humanitarian crisis in Monrovia, but it also corresponds to historic Western tendencies to regard suffering in 'faraway places' as somehow unreal. Susan Sontag notes that the trend often appears in contemporary photojournalism: 'The exhibition in photographs of cruelties inflicted on those with darker complexions in exotic countries continues this offering, oblivious to the considerations that deter such displays of our own victims of violence; for the other, even when not an enemy, is regarded only as someone to be seen, not someone (like us) who also sees. But surely the wounded Taliban soldier begging for his life whose fate was pictured prominently in *The New York Times* also had a wife, children, parents, sisters and brothers, some of who may one day come across

the three color photographs of their husband, father, son, brother being slaughtered – if they have not already seen them.' *Regarding the Pain of Others* (New York: Farrar, Straus, and Giroux, 2003), 72–3.

15 Peter Taylor and Colin Flint, *Political Geography: World-economy, Nation-state and Locality,* 4th ed. (Harlow, Essex: Prentice-Hall, 2000), 54.

16 A Human Rights Watch report notes that the focus of international concern was not Afghanistan's humanitarian crisis, population displacements, famine, and economic ruin: 'instead, several members of the Six Plus Two contact, the six countries bordering Afghanistan, plus Russia and the U.S. that are nominally committed to negotiating an end to the war, are providing military and material support to Afghan parties that have committed gross violations of the laws of war.' Human Rights Watch, 'Afghanistan: Crisis of Impunity, The Role of Pakistan, Russia, and Iran in Fueling the Civil War,' vol. 13, no. 3 (C), July 2001, http://www.hrw.org/reports/2001/afghan2/Afghan0701.htm (accessed 14 August 2001).

17 'President Signs Afghan Women and Children Relief Act, Dec. 21, 2001.' White House Press Release, http://www.whitehouse.gov/news/releases/2001/12/20011212–9.html (accessed 26 March 2005).

18 'Women's Rights in Afghanistan and Beyond,' April Palmerlee, Senior Coordinator for International Women's Issues, Remarks to MSNBC Subscribers, 5 September 2002, http://www.state.gov/g/wi/rls/13593.htm.

19 UN Security Council Resolution 1325 (2000), 31 October 2000.

20 CIA, *The World Fact Book 2002,* http://www.cia.gov/cia/publications/factbook/geos/af.html (accessed 9 August 2003).

21 December 1979 to 5 February 1989.

22 According to the *Annual Report for the Year 2000* by the International Campaign to Ban Landmines, the situation in Afghanistan has improved: 'an estimated five to ten people were injured or killed by mines every day in 1999, compared to an estimated ten to twelve people in 1998 and an estimated twenty to twenty-four people in 1993.'

23 'UNHCR's Contribution to a Special Inter-agency Briefing on Pakistan/Afghanistan – Urgent Need for Camp Sites for Afghan Refugees in Pakistan,' 8 February 2001, http://www.unhcr.ch/news/cupdates/0101afg.htm (accessed 15 August 2001).

24 UNICEF, 'A Humanitarian Appeal for Children and Women, January–December 2001 – Afghanistan,' http://www.unicef.org (accessed 2 January 2002).

25 At 1700 per 100,000 births. Ibid.

26 GDP per capita in 1999 was $800. Ironically, the $2.3 million (1996) external

debt is an economic bright spot. CIA, *The World Factbook 2000*, http://
www.cia.gov/cia/publications/factbook/geos/af.html (accessed
14 August 2001).

27 CIA, *The World Factbook 2000*, http://www.cia.gov/cia/publications/fact-
book/geos/af.html (accessed 14 August 2001).

28 China, Iran, Pakistan, Tajikistan, Turkeminstan, and Uzbekistan plus the
United States and Russia.

29 'A Regional Action Plan adopted by "Six plus Two" Group in New York on
13 September 2000,' UN ODCCP, http://www.odccp.org/uzbekistan/
actionplan.html (accessed 20 August 2001).

30 'Afghanistan Ends Opium Poppy Cultivation,' *ODCCP Update*, June 2001,
http://www.odccp.org/newsletter_2001–06–30_1_page002.html (accessed
20 August 2001).

31 There is much evidence that Taliban were heavily involved in opium cultivation. For example, just before the ban, 'on 17 April [2000] Taliban militia-
men arrested Shujaat Ali Khan, a journalist with the Pakistani daily
Frontier Post, in his hotel room in Kabul. He was taken to a police station
near the presidential palace where he was interrogated at length and
accused of being an American spy. He was eventually released three days
later after a friend intervened on his behalf, but was arrested again by the
Taliban the next day, for no apparent reason. He was kept for nine days in a
filthy cell "crawling with scorpions and all kinds of insects," threatened
with death, questioned daily about his alleged ties with the US Central
Intelligence Agency and accused of manipulating other journalists at the
Frontier Post. Shujaat Ali Khan had come to Afghanistan with a convoy
from the United Nations High Commission for Refugees bringing dozens
of Afghan families from Pakistan. He said he had angered the Taliban by
taking photographs of fields of opium poppies.' Reporters Without Borders, 'Annual Report 2001,' http://www.rsf.fr/uk/home.html (accessed
16 August 2001). Similar evidence has appeared in international press
reports; see Kate Clark, 'The Taleban's Drug Dividend,' *BBC News*, 14 June
2000, http://news.bbc.co.uk/hi/english/world/south_asia/
newsid_783000/7 83268.stm (accessed 26 March 2005).

32 Forcing the international community to recognize the Taliban entailed
obliging it to deal with the regime. In 2001, the Taliban arrested foreign
workers for Relief Now (two Americans, two Australians, and four Ger-
mans) on charges of propagating Christianity, which effectively forced the
three governments concerned to send diplomatic representatives to Kabul.

33 Human Rights Watch, 'Afghanistan: Crisis of Impunity.'

34 The government overthrown by the Taliban, the ISA, was supported by

the United Front. The United Front was mainly made up of the Jamiat-i Islami-yi led by Burhanuddin Rabbani. The United Front's main military leader until his assassination in September 2001 was Ahmad Shad Massoud, the ISA minister of Defense. See HRW, 'Afghanistan: Crisis of Impunity.'

35 In a report in 2000, HRW said 'on April 21, United Front faction Hizb-i Wahdat took control of Bamiyan city, only to lose it after heavy fighting in early May. Following the Hizb-i Wahdat victory, relief workers reported that its forces beat and detained residents suspected of supporting the Taliban, and burned their houses. When Taliban forces retook the city, they reportedly took reprisals by shooting suspected Hizb-i Wahdat supporters, primarily ethnic Shi'a Hazaras, burning hundreds of homes and deporting men to unknown locations.' *HRW World Report 2000*, http://www.hrw.org/wr2k/Asia.htm#Afghanistan (accessed 14 August 2001). HRW also noted that the United Front has targeted specific ethnic groups: 'the United Front has been on the defensive in its home territories, but there have nevertheless been reports of abuses, including summary executions, burning of houses, and looting, principally targeting ethnic Pashtuns and others suspected of supporting the Taliban. The ethnicization of the conflict in the north raises grave concerns about further reprisal attacks on civilians by both sides.' 'Afghanistan: Crisis of Impunity.'

36 Amnesty International, *Annual Report 2001 – Covering Events from January–December 2000 Afghanistan*, http://www.amnesty.org.

37 Under the Taliban, men were hanged, women stoned, and homosexuals crushed beneath collapsing walls. See ibid.

38 See *The World Factbook 2000*.

39 Reporters Without Borders, 'Annual Report 2001.'

40 See 'Taleban Smash TVs,' BBC News, 30 July 1998, http://news.bbc.co.uk/hi/english/world/south_asia/newsid_142000/1 42352.stm (accessed 26 March 2005).

41 'Taleban Rules Out Lifting TV Ban,' BBC News, 23 July 2000, http://news.bbc.co.uk/hi/english/world/south_asia/newsid_847000/8 47408.stm (accessed 26 March 2005).

42 The UF claimed that 30,000 people watched the station each evening.

43 Reporters Without Borders, 'Annual Report 2001.'

44 Ibid. BBC correspondent Owen Bennett-Jones provided glimpses into Afghan life from Pakistan. He reported on a Pakistani soccer team in Kandahar arrested for wearing shorts during a match, which Taliban officials said violated the Islamic dress code. Twelve players were arrested and their heads shaved. Five other players managed to escape. 'Football Tour Cut

Short,' BBC News, 17 July 2000. http://news.bbc.co.uk/1/hi/world/ south_asia/837334.htm (accessed 26 March 2005).

45 http://www.afghan-ie.com.

46 http://www.Taliban.com.

47 Accessed 11 September 2001.

48 Activities included distributing food, blankets, and medicine in refugee camps; organizing home-based classes for women in Afghanistan; attempts to foster global awareness of the issue via the UNCHR; visits to Washington, DC, New York City, Canada, and Japan; and interviews with beggars and prostitutes in Kabul.

49 The website has received three awards: 'Best of '97' at ZDNet http:// www.zdnet.com; 'Hot Site,' http://www.stpt.com; and 'Political Site of the Day,' http://www.aboutpolitics.com.

50 See Michael Dartnell, 'Insurgency Online: Elements for a Theory of Anti-government Internet Communication,' in *Small Wars and Insurgencies* 10, no. 3 (Winter 1999): 117–36.

51 'About RAWA ... ,' http://www.rawa.org (accessed 22 August 2001).

52 Ibid.

53 See Amnesty International, http://www.amnesty.se/women/23ea.htm.

54 See http://rawa.fancymarketing.net/goals.htm (accessed 25 August 2001).

55 See http://pz.rawa.org.

56 In August 2001, both of these publications, which sell for about $10 U.S. each, were listed as out of stock at http://songs.rawa.org/rawa/burst.htm.

57 The translated reports, available at http://songs.rawa.org/rawa/ai.htm, are: 'Afghanistan: International Responsibility for Human Rights Disaster'; 'Afghanistan: Grave Abuses in the Name of Religion'; 'Afghanistan: Cruel, Inhuman or Degrading Treatment or Punishment'; 'Human Rights Defenders in Afghanistan: Civil Society Destroyed'; 'Afghanistan: The Human Rights of Minorities'; 'Women in Afghanistan: Pawns in Men's Power Struggles'; 'Children Devastated by War: Afghanistan's Lost Generations'; and 'Refugees from Afghanistan: The World's Single Largest Refugee Group.' A link to http://www.amnesty.org is also provided.

58 The documents here point out that, even prior to international outrage over destruction of ancient Buddhist sites in 2001, the conflict had already resulted in the destruction of Kabul's renowned national museum and the sale of many surviving priceless artifacts abroad.

59 The restrictions are found at http://songs.rawa.org/rawa/rules.htm. RAWA has provided translations in Spanish, Italian, French, and German. See Appendix 1 below.

60 RAWA has reproduced the report at http://songs.rawa.org/rawa/channe14.htm.

61 In 'RAWA Holds Press Conference after Being Forced to Cancel Its Protest Rally April 27, 2001' at http://songs.rawa.org/rawa/apr28–01r.htm, RAWA stated : 'Every year on the 28th of April, the Revolutionary Association of the Women of Afghanistan (RAWA) holds a public demonstration to mark the day that fundamentalists took power in Afghanistan in 1992. This year, police in Peshawar stopped RAWA from demonstration and informed us that we would demonstrate at our own risk and should not count on police protection even if attacked. Also the fundamentalist groups who attacked RAWA rally on Dec. 10 last year, threatened RAWA, saying that if we hold demonstration, they will attack again. Since a large faction of protesters could not reach Pakistan from Afghanistan and we were denied permission by police, RAWA was forced to cancel the demonstration. A press conference was held instead on 27th of April in Peshawar to condemn the black day and the Pakistani government for stopping RAWA to demonstrate against Taliban and Jehadis in Peshawar.'

62 See http://rawa3.hackmare.com/rawa/movies1.html.

63 See http://www.jubillenium.com/.

64 See http://www.amnesty.org/ailib/countries/indx311.htm.

65 See http://www.unhcr.ch/refworld/un/chr/chr95/country/64.htm.

66 See http://www.femaid.org/.

67 See http://www.femaid.org/RAWA.html.

68 See http://www.afghanradio.com/azadi.html.

69 See http://www.womenasia.com/W4Wafghan.

70 See http://www.asap-net.org/.

71 See http://www.octavesbeyondsilence.com/.

72 See http://afghanwomensmission.org/index.shtml.

73 See http://www.homa.org/mission.html.

74 See http://www.afghans.bit.com.au/.

75 See http://www.SearchShack.com/.

76 See http://www.now.org. At the time the site was visited in August 2001, no items relating to Afghan women were posted.

77 See http://www.feminist.org/afghan/intro.asp.

78 See http://www.golshan.com/.

79 See http://rawa.fancymarketing.net/donation.htm. In August 2003, the donations were being accepted in British pounds and Euros.

80 See http://rawa.fancymarketing.net/help.htm.

81 See http://rawa.fancymarketing.net/petition.htm.

82 See *North County Times*, 19 December 2000, http://www.nctimes.com/news/121900/y.html.

83 See Nicholas Negroponte, *Being Digital* (New York: Vintage, 1995).

84 Wilkinson, a prominent analyst of terrorism, argued in the 1970s that 'the crucial advantage of a liberal state to the terrorist, however, is the freedom of the media. The terrorist operating within such a society knows that his acts of terrorism will be instantly publicized by the television, radio and Press and that pictures of a really sensational attack or outrage can be relayed round the world with the aid of T.V. communications satellites,' *Terrorism and the Liberal State* (London: Macmillan, 1977), 103.

85 Echoing Wilkinson, Hoffman argues that cable network news has become 'not just an "opinion shaper" but a "policy driver," its presenters and on-air analysts racing to define the range of options at a government's disposal or interpret likely public reaction – and its repercussions.' *Inside Terrorism* (London: Victor Gollancz, 1998), 151.

4 Insurgency Online as Media Relay: The Web Activism of the MRTA

1 An early version of this chapter was presented as 'Insurgency Online: http:// burn.ucsd.edu/~ats/mrta.htm' at the Comparative Politics (Developing) session of Canadian Political Science Association (CPSA) Annual Meeting at the University of Ottawa, Ottawa, Ontario, on 2 June 1998. A revised version was published as 'Insurgency Online: Elements for a Theory of Anti-government Internet Communication' in *Small Wars and Insurgencies* 10, no. 3 (Winter 1999): 117–36.

2 Mark Ward, 'Websites Spread al-Qaeda Message,' BBC News, 12 December 2002, http://news.bbc.co.uk/2/hi/technology/2566527.stm (accessed 13 December 2002).

3 See Nick Stevenson, *The Transformation of the Media: Globalisation, Morality and Ethics* (London: Longman, 1999), and Dan Schiller, 'How to Think about Information,' in *The Political Economy of Information*, ed. Vincent Mosco and Janet Wasko (Madison: University of Wisconsin Press, 1988), 27–43.

4 See, for example, the work of Jürgen Habermas and Raymond Williams.

5 Douglas Kellner, 'Intellectuals, the New Public Spheres, and Techno-Politics,' in *The Politics of Cyberspace*, ed. Chris Toulouse (London: Routledge, 1998), 174.

6 Mark D. Alleyne, *International Power and International Communication* (London: Macmillan, 1995), 3.

7 Schiller, 'How to Think about Information,' 41.

8 Examples of anti-government/insurgent groups that have websites include

Abkhazian separatists, the Animal Liberation Front, the Algerian Islamic
Salvation Front, the National Union for the Total Independence of Angola
(UNITA), the Free Burma Coalition, the Centre of Human Rights and Dem-
ocratic Development in China, Columbia's Ejercito de Liberacion Nacional
(ELN), the Cuban dissident movement, the East Timor Action Network, the
Basque Herri Batasuna, Mexico's EZLN (Zapatista National Liberation
Army), the Iraqi National Congress, Hezbollah, Japan's AUM Shinrikyo
cult, the Jammu Kashmir Liberation Front (JKLF), and many others.

9 Phil Agre, 'Some Thoughts about Political Activity on the Internet,' http://
www.ucaqld.com.au/news/4political/activity.html, August 1996
(accessed 3 August 1999).

10 One journalist notes that 'the Web offers ... a powerful interactive communications tool that bypasses the editorial control of other mass media and
promises to reach far more eyes than a handful of leaflets passed out on a
street corner.' Matthew McAllester, 'Peruvian Rebels Go to the Web,' *New
York Newsday*, 8 January 1997, http://www.newsday.com.

11 Truth and Reconciliation Commission (TRC), *Final Report*, 28 August 2003,
2.

12 Reporters Without Borders, 'Peru – Annual Report 2002,' http://
www.rsf.fr.print.php3?id_article=1417 (accessed 14 December 2004).

13 Human Rights Watch, 'Human Rights Watch World Report 1998 – Peru,'
http://www.hrw.org/worldreport/Americas-09.htm#P902_I92900
(accessed 14 December 2004).

14 TRC, 19.

15 International Press Institute, 'World Press Freedom Review 1997 – Peru,'
http://www.freemedia.at/wpfr/Americas/peru.htm (accessed 14 December 2004).

16 'Struggle against Neo-Liberalism!' MRTA Communiqué, no date.

17 'Peru Country Report on Human Rights Practices for 1998,' Bureau of
Democracy, Human Rights and Labor, 26 February 1998.

18 'Report of the Special Rapporteur on the independence of judges and lawyers, Mr Param Cumaraswamy: Report on the Mission to Peru,' Office of
the UN High Commissioner for Human Rights, Geneva, Switzerland, 19
February 1998.

19 Although it admitted that certain improvements occurred in 1996, the RSF
report for 1997 cited a number of cases in which journalists were jailed,
arrested, attacked, threatened, and harassed, administrative, legal and economic pressure was applied, and obstacles raised to the domestic free flow
of information. See http://www.calvacom.fr/rsf/RSF_VA/Rapp_VA/
Carte_VA/Rapp_VA.html.

20 McAllester, 'Peruvian Rebels Go to the Web.'
21 Tom Vogel, Matt Moffett, and Jed Sandberg, 'Tupac Amaru's Web Page Is Hot Spot on the Internet,' *Wall Street Journal*, 6 January 1997.
22 The U.S. State Department publishes a yearly list of groups that it labels 'terrorists.' By August 2003, the MRTA no longer appeared on the list. However, on 5 December 2001 Secretary of State Colin Powell placed the following groups on the Terrorist Exclusion List: Al-Ittihad al-Islami (AIAI); Al-Wafa al-Igatha al-Islamia; Asbat al-Ansar; Darkazanli Company; Salafist Group for Call and Combat (GSPC); Islamic Army of Aden; Libyan Islamic Fighting Group; Makhtab al-Khidmat; Al-Hamati Sweets Bakeries; Al-Nur Honey Center; Al-Rashid Trust; Al-Shifa Honey Press for Industry and Commerce; Jaysh-e-Mohammed; Jamiat al-Ta'awun al-Islamiyya; Alex Boncayao Brigade (ABB); Army for the Liberation of Rwanda (ALIR) – aka: Interahamwe, Former Armed Forces (EX-FAR); First of October Antifascist Resistance Group (GRAPO) – aka: Grupo de Resistencia Anti-Fascista Premero De Octubre; Lashkar-e-Tayyiba (LT) – aka: Army of the Righteous; Continuity Irish Republican Army (CIRA) – aka: Continuity Army Council; Orange Volunteers (OV); Red Hand Defenders (RHD); New People's Army (NPA); People Against Gangsterism and Drugs (PAGAD); Revolutionary United Front (RUF); Al-Ma'unah; Jayshullah; Black Star; Anarchist Faction for Overthrow; Red Brigades-Combatant Communist Party (BR-PCC); Revolutionary Proletarian Nucleus; Turkish Hizballah; Jerusalem Warriors; Islamic Renewal and Reform Organization; The Pentagon Gang; Japanese Red Army (JRA); Jamiat ul-Mujahideen (JUM); Harakat ul Jihad i Islami (HUJI); The Allied Democratic Forces (ADF); and The Lord's Resistance Army (LRA). The following additional groups were added to the list on 18 February 2003: Al Taqwa Trade, Property and Industry Company Ltd. (fka Al Taqwa Trade, Property and Industry; fka Al Taqwa Trade, Property and Industry Establishment; fka Himmat Establishment); Bank Al Taqwa Ltd. (aka Al Taqwa Bank; aka Bank Al Taqwa); Nada Management Organization (fka Al Taqwa Management Organization SA); Youssef M. Nada & Co. Gesellschaft M.B.H.; Ummah Tameer E-Nau (UTN) (aka Foundation for Construction; aka Nation Building; aka Reconstruction Foundation; aka Reconstruction of the Islamic Community; aka Reconstruction of the Muslim Ummah; aka Ummah Tameer I-Nau; aka Ummah Tamir E-Nau; aka Ummah Tamir I-Nau; aka Ummat Tamir E-Nau; aka Ummat Tamir-I-Pau); Loyalist Volunteer Force (LVF); Ulster Defense Association (aka Ulster Freedom Fighters); Afghan Support Committee (aka Ahya ul Turas; aka Jamiat Ayat-ur-Rhas al Islamia; aka Jamiat Ihya ul Turath al Islamia; aka Lajnat el Masa Eidatul Afghania); and Revival of Islamic Heritage Society (Pakistan

and Afghanistan offices – Kuwait office not designated) (aka Jamia Ihya ul Turath; aka Jamiat Ihia Al- Turath Al-Islamiya; aka Revival of Islamic Society Heritage on the African).

23 See, for example, Amit Asaravala, 'College Questioning Site's Link,' *Wired News*, 28 September 2002, http://www.wired.com/news/politics/ 0,1283,55450,00.html (accessed 18 August 2003).

24 A mirror website, Vox Rebelde, is the group's main Spanish-language site.

25 Throughout the Lima hostage crisis, the MRTA Solidarity Page posted interviews on events. Interviews available on 22 April 1997 included: 'What Are the Goals of the Embassy Occupation?' (19 December 1996); 'How Long Will the Residence Stay Occupied?' (24 December 1996); 'Is the MRTA's Action Weakening Fujimori?' (30 December 1996); and 'Is a Solution in Lima at Hand?' (24 March 1997).

26 On 22 April 1997, these texts included: 'Interview with Victor Polay – 1990'; 'Interview with an MRTA Leader, Comandante Andres – January 1991'; 'The Situation of MRTA Political Prisoners in Peru – May 1996'; 'Letter from MRTA Political Prisoners – November 1996'; 'The Lives of Political Prisoners in Peru Are in Danger!'; and 'Neo-Liberalism and Globalization.'

27 The latter are members of the DHKP-C (Revolutionary People's Liberation Party-Front). The statement also included email and Web addresses: DHKC Informationbureau Amsterdam: dhkc@ozgurluk.xs4all.nl; 'Classwar in Turkey and Kurdistan,' http://www.xs4all.nl/~ozgurluk; 'Turkey Contra-Guerrilla State,' http://www.xs4all.nl/~ozgurluk/pub/contrind.html; 'Turkey Mailinglist Mirror,' http://www.xs4all.nl/~ozgurluk/ml.html; 'KURTULUS HAFTALIK SIYASI GAZETE,' http://www.kurtulus.com.

28 See Nicaragua Solidarity Network of Greater New York, 339 Lafayette St., New York, NY 10012 USA, Tel: (212) 674–9499; Email: nica-net@nyxfer.blythe.org. Web: home.earthlink.net/nicadlu/nsnhome.html (accessed 26 March 2005).

29 The Italian Solidarity Page 'Solidarietà al M.R.T.A.' includes high-quality graphics and links to other left and extreme-left websites. Links on this page included the MRTA mirror site 'Voz Rebelde,' the Lima newspaper *La Republica*, CNN World News, Yahoo!, Reuters International Summary, and the Tactical Media Crew Home Page homepage and email address, tmcrew@mail.nexus.it.

30 See http://users.cybercity.dk/ ccc17427/ (accessed 22 April 1997).

31 Vogel, Moffett, and Sandberg, 'Tupac Amaru's Web Page Is Hot Spot on the Internet.'

32 The dates for hits are my own and were gathered through visiting the site.

33 In 1999, RSF singled out twenty states, so-called 'enemies of the Net,' that severely restrict or altogether curtail access to the Internet by their citizens: Azerbaijan, Belarus, Burma, China, Cuba, Iran, Iraq, Kazakhstan, Kirghizia, Libya, North Korea, Saudi Arabia, Sierra Leone, Sudan, Syria, Tajikistan, Tunisia, Turkmenistan, Uzbekistan, and Vietnam. See http://www.calvacom.fr/rsf/RSF_VA/Rapp_VA/Carte_VA/Rapp_VA.html. A number of states in this list have close political, military, and/or economic ties to the United States.

34 'Peruvian Rebels Wage Propaganda War on Internet,' Reuters, 3 January 1997.

35 T. Thorton, 'Terror as a Weapon of Political Agitation,' in *Internal War: Problems and Approaches*, ed. Harry Eckstein (New York: Free Press, 1964), 78–82.

36 Max Weber, *The Methodology of Social Sciences* (New York: Free Press, 1949), 110.

37 Clifford Geertz, *The Interpretation of Cultures* (New York: Basic Books, 1973), 316.

38 My book *Action directe: Ultra-left Terrorism in France, 1979–1987* (London: Frank Cass, 1995), discusses in more depth the conceptual barriers to the analysis of terrorism by non-state actors.

39 See Dartnell, *Action Directe*.

40 Interactivity enables Internet users to alter certain aspects of their environment. It is a method of control and contingent response between user and medium. Terms to describe interactive systems include 'multimedia,' 'hypermedia,' 'infotainment,' and 'edutainment.'

41 The initial demographic was especially characterized by 18–35-year-old, computer-literate, white, English-speaking North American men.

42 Mark S. Bonchek, 'Grassroots in Cyberspace: Using Computer Networks to Facilitate Political Participation,' MIT Artificial Intelligence Laboratory Working Paper 95–2.2, presented at the 53rd Annual Meeting of the Midwest Political Science Association, Chicago, IL, 6 April 1995, http://www.ai.mit.edu/people/msb/pubs/grassroots.html.

43 James E. Katz and Ronald E. Rice, *Social Consequences of Internet Use: Access, Involvement, and Interaction* (Cambridge, MA: MIT Press, 2002), 322–3.

44 Edwyn James, 'Learning to Bridge the Digital Divide,' *OECD Observer*, 14 January 2001.

45 My own experience in doing research on the French extreme-left terrorist group *Action directe* (AD) is a case in point. When I started the research, my overriding concern was *how* to get information on an illegal organization about which few people knew or cared to know anything. It took

several months and considerable luck to assemble my research. The same difficulty would not exist today since groups such as AD now routinely place information about themselves and their activities on the Web.

46 'Peruvian Rebels Wage Propaganda War on Internet.'

47 The journalists were Cecilia Valenzuela, former editor of the television program *Aqui y ahora*; Luis Ibérico, former presenter of the program *Contrapunto*; Baruch Ivcher, majority shareholder in *Frecuencia Latina*; Gustavo Mohme, Angel Páez, and Edmundo Cruz, publisher and journalists for *La República*; and Fernando Respigliosi, from the magazine *Caretas*. Reporters Without Borders, 'The Enemies of the Internet – Peru,' http://www.rsf.fr/rsf/uk/html/internet/pays_internet/perou.html (accessed 14 December 2004).

48 Alberto Manguel, *Reading Pictures: A History of Love and Hate* (New York: Random House, 2000), 71.

Conclusion: Web Activism – A Messenger That Shapes Perceptions

1 Bertolt Brecht, 'The Radio as an Apparatus of Communication,' *Brecht on Theatre*, trans. John Willett (New York: Hill and Wang, 1977), 52.

2 Walter Benjamin, 'The Work of Art in the Age of Mechanical Reproduction,' in *Illuminations* (London: Fontana, 1977), 226.

3 Ibid., 244.

4 See Jamie Frederic Metzl, 'Rwandan Genocide and the International Law of Radio Jamming,' *American Journal of International Law* 91, no. 4 (October 1997): 628, http://www.asil.org/ajil/radio.htm (accessed 23 August 2003). Metzl notes that in spite of 'the inadequacies of Rwanda's information infrastructure, the killings were carried out in a highly systematic and synchronized manner, the result of careful advance planning. One central feature of this planning was the use of radio, particularly the semiprivate station, *Radio-Television Libre des Milles Collines* (RTLM). RTLM was founded by leading Hutu extremists in the Rwandese Government in mid-1993 in response to reforms that had allowed moderates to take positions in the administration, including the Ministry of Information, which controlled Radio Rwanda, the official governmental station. Although established as a jointly funded stock company, the station was essentially the tool of Hutu extremists from the government, military and business communities.'

5 Paul Virilio, *Open Sky*, trans. Julie Rose (London: Verso, 1997), 71.

6 Given that the perpetrator of these attacks was a U.S. citizen, the incident simultaneously points to the globalizing and privatizing impacts of IT on

security. See 'Al-Jazeera Hacker Pleads Guilty,' BBC News, 13 June 2003, http://news.bbc.co.uk/2/hi/americas/2987342.stm (accessed 25 July 2003).

7 Graham Meikle, *Future Active: Media Activism and the Internet* (New York: Routledge, 2002), 26.

8 Meikle defines hacktivism as electronic 'political resistance' rather than terrorism. Ibid., 141.

9 Mark Ward, 'Websites Spread al-Qaeda Message,' BBC News, 12 December 2002, http://news.bbc.co.uk/2/hi/technology/2566527.stm (accessed 13 December 2002).

10 Shockwave is a standard software technology available at http://www.macromedia.com.

11 The URL for the website that posted the file is http://www.citylinkcomputers.com.

12 I would like to thank Mr Mohammed Zigby, a PhD candidate in Islamic Studies at McGill University, for his assistance in examining the multimedia file of the *du'a* and verifying the contents of the audio recording.

13 For a transcript of the *du'a*, see Appendix 3, 'Translation of the *Du'a* of Sheikh Muhammed Al Mohaisany.'

14 All references to the *du'a* are from the version included in its entirety in Appendix 3.

15 Benjamin stated that 'photographs become the standard evidence for historical occurrences, and acquire a hidden political significance. They demand a specific kind of approach; free-floating contemplation is not appropriate to them. They stir the viewer; he feels challenged by them in a new way. At the same time picture magazines begin to put up signposts for him, right ones or wrong ones, no matter. For the first time, captions have become obligatory. And it is clear that they have an altogether different character than the title of a painting.' 'The Work of Art in the Age of Mechanical Reproduction,' 228.

16 James Rosenau and David Johnson, 'Information Technologies and Turbulence in World Politics,' in *Technology, Development, and Democracy: International Conflict and Cooperation in the Information Age*, ed. Julianne Emmons Allison (Albany: SUNY Press, 2002), 74.

17 For a discussion of the interaction of caption/text and photograph, see Susan Sontag, *On Photography* (New York: Picador, 2001), 107–8.

18 John Taylor, *Body Horror: Photojournalism, Catastrophe and War* (New York: New York University Press, 1998), 64.

19 See Marie-Laure Ryan, *Narrative across Media: The Languages of Storytelling* (Lincoln: University of Nebraska Press, 2004).

20 Barbie Zelizer, *Remembering to Forget: Holocaust Memory through the Camera's Eye* (Chicago: University of Chicago Press, 1998), 215.

21 These frontiers are seen, for example, in the fortresses built at Lille in the north and Bellegarde in Rousillon.

22 In the late 1990s and early 2000s, India and Pakistan engaged in a series of border clashes over the disputed region of Kashmir. Although conflict between Hindus and Muslims preceded the European intrusion into South Asia, that event facilitated its re-embodiment as a war of nationalisms (Indian and Pakistani).

23 This discussion will refer to globalization as defined by Held, McGrew, Goldblatt, and Perraton in *Global Transformations* as 'a process (or set of processes) which embodies a transformation in the spatial organization of social relations and transaction – assessed in terms of their extensity, intensity, velocity and impact – generating transcontinental or inter-regional flows and networks of activity, interaction, and the exercise of power' (16).

24 Frederick, *Global Communication and International Relations*, 16.

25 For a discussion on changing ideas of time and space see James Anderson's work on the shifting notion of territoriality as it pertains to Northern Ireland, *In Search of Ireland: A Cultural Geography* (London: Routledge, 1997). See also: Darin Barney, *Prometheus Wired: The Hope for Democracy in the Age of Network Technology* (Vancouver: UBC Press, 2000); Geert Lovnik, *Dark Fiber: Tracking Critical Internet Culture* (Cambridge, MA: MIT Press. 2002); Lev Manovich, *The Language of New Media* (Cambridge, MA: MIT Press, 2001); Brian Massumi, *Parables for the Virtual: Movement, Affect, Sensation* (Durham, NC: Duke University Press, 2002); and David Norman Rodowick, *Reading the Figural, or, Philosophy after the New Media* (Durham, NC: Duke University Press, 2001).

26 The Italian philosopher Giorgio Agamben refers to this as the state's power over 'bare life.' *Homo Sacer: Sovereign Power and Bare Life* (Stanford, CA: Stanford University Press, 1998).

27 David Held, 'Violence, Law and Justice in a Global Age,' *SSRC after September 11 Archive*, 5 November 2001, http://www.ssrc.org/sept11/essays/held.htm (accessed 29 December 2002).

28 Sontag, *On Photography*, 23.

29 Paul Dourish, *Where the Action Is: The Foundations of Embedded Interaction* (Cambridge, MA: The MIT Press, 2001), 2.

30 For a discussion of the complex impacts of television and globalization, see Lisa Parks and Shanti Kumar, eds, *Planet TV: A Global Television Reader* (New York: New York University Press, 2003).

31 James Rosenau, *Turbulence in World Politics: A Theory of Change and Continuity* (Princeton, NJ: Princeton University Press, 1990), 11.

32 See, for example, Roger Fenton and James Robertson's photographs of the Crimean War, Felice Beato's images of the 1860 Anglo-French expeditionary force that invaded China, or the American Civil War photographs of Alexander Gardner, James Gibson, George Barnard, Andrew Russell, Timothy O'Sullivan, and Thomas Roche in Therese Mulligan and David Wooters, eds, *1000 Photo Icons: George Eastman House* (Cologne: Taschen 2002), 250–71.

33 Lisa Parks, *'Our World*, Satellite Televisuality, and the Fantasy of Global Presence,' *Planet TV*, 89.

34 See, for example, Yuri the Yaba, 'Yabasta: Mobilizing Global Citizenship through Mass Direct Action,' in *Global Uprising: Confronting the Tyrannies of the 21st Century*, ed. Neva Welton and Linda Wolf (Gabriola Island, BC: New Society Publishers, 2001), 53–6. Yabasta articulates a vision of post-territorial politics that is 'post-ideological, and that's its innovative potential: it privileges strategies and specific actions, rather than affiliating with a historically burdened tradition. Yabasta practices direct action without sectarian divisions and does not try to decide what kind of change each community may want – they can organize and decide for themselves. It has no formal structure; we dress up to protect and support each other to become a critical mass' (56).

35 See 'What the World Thinks in 2002: How Global Publics View: Their Lives, Their Countries, the World, America,' *The Pew Global Attitudes Survey*, The Pew Research Center, 4 December 2002, http://people-press.org/reports/display.php3?ReportID=165 (accessed 30 December 2002).

36 Robert Burnett and P. David Marshall, *Web Theory: An Introduction* (London: Routledge, 2003), 61.

Bibliography

Books and Articles

Adatto, Kiku. *Picture Perfect: The Art and Artifice of Public Image Making*. New York: Basic Books, 1993.

'After the Cold War in the Wake of the Terror.' *NACLA Report on the Americas* 35, no. 3 (November-December 2001).

Agamben, Giorgio. *Homo Sacer: Sovereign Power and Bare Life*. Stanford, CA: Stanford University Press, 1998.

'Ah, Peace.' *The Economist*, 28 October 1995, 38.

Alexander, Catherine. 'National Security in a Wired World.' In *Digital Democracy: Policy and Politics in the Wired World*, edited by Cynthia Alexander and Leslie Pal, 46–62. Toronto: Oxford University Press, 1998.

Alexander, Cynthia J., and Leslie A. Pal. *Digital Democracy: Policy and Politics in the Wired World*. Toronto: Oxford University Press, 1998.

Alleyne, Mark D. *International Power and International Communication*. London: Macmillan, 1995.

Allison, Julianne Emmons, ed. *Technology, Development, and Democracy: International Conflict and Cooperation in the Information Age*. Albany: SUNY Press, 2002.

Anderson, Benedict. *Imagined Communities: Reflections on the Origin and Spread of Nationalism*. London: Verso, 1983.

Anderson, James. *In Search of Ireland: A Cultural Geography*. London: Routledge, 1997.

Anderson, Peter J. *The Global Politics of Power, Justice and Death*. London: Routledge, 1996.

Appadurai, Arjun. *Modernity at Large: Cultural Dimensions of Globalization*. Minneapolis: University of Minnesota, 1996.

– 'Disjuncture and Difference in the Global Cultural Economy.' In *Planet TV: A Global Television Reader*, 41–52. New York: New York University Press, 2003.

Aristotle. *The Politics*, edited by Ernest Barker. London: Oxford University Press, 1974.

Armstrong, Sally. *Veiled Threat: The Hidden Power of the Women of Afghanistan.* Toronto: Penguin, 2002.

Arquilla, John, and David Ronfeldt. 'Cyberwar Is Coming!' *Comparative Strategy* 12, no. 2 (Spring 1993): 141–65.

– *The Advent of Netwar.* Santa Monica, CA: Rand, 1996.

– *In Athena's Camp: Preparing for Conflict in the Information Age.* Santa Monica, CA: RAND, 1997.

Arquilla, John, David Ronfeldt, Graham Fuller, and Melissa Fuller. *The Zapatista Social Netwar in Mexico.* Santa Monica, CA: RAND, 1998.

Article 19 Organization. *Broadcasting Genocide: Censorship, Propaganda and State Sponsored Violence in Rwanda 1990–1994* (Article 19, 1996).

Aughey, A., and D. Morrow, eds. *Northern Ireland Politics.* London: Longman, 1996.

Azizy, Fahimam. *Gender Integration and Women in Co-Operative Development Country Study of Afghanistan.* New Delhi: International Co-operative Alliance, Regional Office, 1992.

'Back to War in Afghanistan.' *The Economist*, 31 July 1999, 33.

Baklarz, Ronald, and Richard Forno. *The Art of Information Warfare: Insight into the Knowledge Warrior Philosophy*, 2nd ed. Universal Publishers, 1999.

Bakshi, G.D. *Afghanistan: The First Faultline War.* New Delhi: Lancer Publishers & Distributors, 1999.

Barney, Darin. *Prometheus Wired: The Hope for Democracy in the Age of Network Technology.* Vancouver: UBC Press, 2000.

'The Battle for Afghanistan.' *The Economist*, 31 May 1997, 37–8.

Baudrillard, Jean. *The Spirit of Terrorism.* Translated by Chris Turner. London: Verso, 2002.

Beier, J. Marshall, and Steven Mataija, eds. *Cyberspace and Outer Space: Transitional Challenges for Multilateral Verification in the 21st Century. Proceedings of the Fourteenth Annual Ottawa NACD Verification Symposium.* Toronto: Centre for International and Security Studies, York University, 1997.

Bell, David V.J. 'Global Communications, Culture, and Values: Implications for Global Security.' In *Building a New Global Order: Emerging Trends in International Security*, edited by David Dewitt, David Haglund, and John Kirton, 159–84. Toronto: Oxford University Press, 1993.

Bellow, H.W. *An Enquiry into the Ethnography of Afghanistan.* New Delhi: Bhavana Books & Prints, 2001.

Benjamin, Walter. 'The Work of Art in the Age of Mechanical Reproduction.' In
    Illuminations, 219–53. London: Fontana, 1977.
– 'The Author as Producer' In *Reflections*, 220–38. London: Harcourt Brace
    Jovanovich, 1978.
Berger, Pamela. 'The Historical, the Sacred, the Romantic: Medieval Texts into
    Irish Watercolors.' In *Visualizing Ireland: National Identity and the Pictorial Tra-
    dition*, edited by Adele Dalsimer, 71–87. Boston: Faber and Faber, 1993.
Berners-Lee, T., and M. Fischetti. *Weaving the Web: The Original Design and Ulti-
    mate Destiny of the World Wide Web by Its Inventor.* New York: Harper San
    Francisco, 1999.
Bolter, Jay David, and Diane Gromala. *Windows and Mirrors: Interaction Design,
    Digital Art, and the Myth of Transparency.* Cambridge, MA: MIT Press, 2003.
Bolter, Jay David, and Richard Grusin. *Remediation: Understanding New Media.*
    Cambridge, MA: MIT Press, 1999.
Brecht, Bertolt. 'The Radio as an Apparatus of Communication.' In *Brecht on
    Theatre*, translated by John Willett, 51–3. New York: Hill and Wang, 1977.
Brentjes, Burchard, and Helga Brentjes. *Taliban: A Shadow over Afghanistan.*
    Varanasi: Rishi Publications, 2000.
Brett, Sebastian. 'Peru Confronts a Violent Past: The Truth Commission Hear-
    ings in Ayacucho.' Human Rights Watch (2003).
Brewer, John. *Party Ideology and Popular Politics at the Accession of George III.*
    Cambridge: Cambridge University Press, 1981.
'Brutal Winter.' *The Economist*, 16 December 1995, 35.
Burnett, Robert, and P. David Marshall. *Web Theory: An Introduction.* London:
    Routledge, 2003.
Byman, Daniell, and Zalmay Khalilzad. 'Afghanistan: The Consolidation of a
    Rogue State.' *Washington Quarterly* 23, no. 1 (Winter 2000): 65–7.
Cairncross, F. *The Death of Distance: How the Communications Revolution Will
    Change Our Lives.* Boston: Harvard Business School Press, 1997.
Cameron, Maxwell A., Philip Mauceri, and Cynthia McClintock. *The Peruvian
    Labyrinth: Polity, Society, Economy.* University Park: Pennsylvania State Uni-
    versity Press, 1997.
Castells, Manuel. *The Rise of the Network Society.* Malden, MA: Blackwell,
    1996.
'Celebrating 25 Years of Networking.' *Women in Action* 2 (1999): 75.
Chopra, V.D., ed. *Afghanistan and Asian Stability.* New Delhi: Gyan Publishing
    House, 1998.
Chopra V.D. and Rasgotra, M., eds. *Genesis of Regional Conflicts – Kashmir,
    Afghanistan, Cambodia,West Asia, Chechnya.* New Delhi: Gyan Publishing
    House, 1995.

Chrétien, Jean-Pierre, et al., eds. *Rwanda: Les medias du genocide*. Paris: Karthala, 1995.

Cohen, Anthony. *The Symbolic Construction of Community.* New York: Routledge, 1985.

Comor, Edward. *Communication, Commerce and Power: The Political Economy of America and the Direct Broadcast Satellite, 1960–2000.* London and New York: Macmillan and St Martin's Press, 1998.

– 'Governance and the Nation-State in a Knowledge-Based Political Economy.' In *Approaches to Global Governance Theory*, edited by Timothy J. Sinclair and Martin Hewson, 117–34. Albany: SUNY Press, 1999.

– 'The Role of Communication in Global Civil Society: Forces, Processes, Prospects.' *International Studies Quarterly* 45, no. 3 (September 2001): 309–23.

– 'New Technologies and Consumption: Contradictions in the Emerging World Order.' In *Information Technologies and Global Politics: The Changing Scope of Power and Governance*, edited by James Rosenau and J.P. Singh, 169–85. Albany: SUNY Press, 2002.

– ed. *The Global Political Economy of Communication: Hegemony, Telecommunication and Information Economy.* New York: Palgrave Macmillan, 1996.

Cooper, Jeffrey R., and Christopher Burton. 'The Information Revolution, the Military, and Arms Control.' In *Cyberspace and Outer Space: Transitional Challenges for Multilateral Verification in the 21st Century*, edited by J. Marshall Beier and Steven Mataija, 93–104. Proceedings of the Fourteenth Annual Ottawa NACD Verification Symposium. Toronto: Centre for International and Security Studies, York University, 1997.

'Cops: The Rise of Crime, Disorder and Authoritarian Policing.' *NACLA Report on the Americas* 37, no. 2 (September-October 2003).

Cowhey, Peter. 'The International Telecommunications Regime: The Political Roots of Regimes for High Technology.' *International Organization* 44, no. 2 (1990): 169–200.

Crandall, Robert, and Robert Flamm, eds. *Changing the Rules: Technological Change, International Competition, and Regulation in Communication.* Washington, DC: Brookings Institution, 1989.

Curtin, Nancy J. *The United Irishmen: Popular Politics in Ulster and Dublin 1791–1798.* Oxford: Clarendon Press, 1998.

Dartnell, Michael Y. 'France's *Action directe*: Terrorists in Search of a Revolution.' In *European Terrorism*, edited by Edward Moxon-Browne, 187–218. New York: G.K. Hall, 1994.

– *Action Directe: Ultra-left Terrorism in France, 1979–1987.* London: Frank Cass, 1995.

- 'Action Directe.' In Encyclopedia of World Terrorism, edited by M. Crenshaw and J. Pimlott, 557–9. New York: M.E. Sharpe, 1996.
- 'Where Do Angels Hang in the Cybernet Nineties? – Meditations on Theological Politics.' In Digital Delirium, 219–31. New York and Montreal: St Martin's Press/New World Perspectives, 1997.
- 'HTML (HyperText Markup Language) as Needlepoint.' CTheory 22, no. 3, article 77 (10 November 1999).
- 'Insurgency Online: Elements for a Theory of Anti-government Internet Communication.' Small Wars and Insurgencies 10, no. 3 (Winter 1999): 117–36.
- 'A Legal Inter-Network for Terrorism: Issues of Globalization, Fragmentation and Legitimacy.' Terrorism and Political Violence 11, no. 4 (Winter 1999): 197–208.
- 'The Belfast Agreement: Peace Process, Europeanization and Identity.' Canadian Journal of Irish Studies 26, no. 1 (2001).
- 'The Electronic Starry Plough: The Enationalism of the Irish Republican Socialist Movement (IRSM).' First Monday 6, no. 12 (December 2001).
- 'Hyperterrorism: A New Form of Globalized Conflict.' CTheory (26 September 2001).
- 'Information Technology and the Web Activism of the Revolutionary Association of the Women of Afghanistan: Electronic Politics and New Global Conflict.' In Bombs and Bandwidth: The Emerging Relationship between Information Technology and Security, edited by Robert Latham, 251–67. New York: New Press, 2003.
- 'Multimedia as Transgressive Practice: The D'ua of Sheikh Muhammed Al Mohaisany.' In Governance and Global (Dis)orders: Trends Transformations and Impasses. Selected Proceedings of the Eleventh Annual Conference of the Centre for International and Security Studies in Conjunction with the Fourth Annual Conference of the Nathanson Centre for the Study of Organized Crime and Corruption, edited by Alison Howell, 289–312. Toronto: York Centre for International and Security Studies, 2004.
- Debrix, François, and Cynthia Weber, eds. Rituals of Mediation: International Politics and Social Meaning. Minneapolis: University of Minnesota Press, 2003.
- Deibert, Ronald. Parchment, Printing, and Hypermedia. New York: Columbia University Press, 1997.
- 'Altered Worlds: Social Forces in the Hypermedia Environment.' In Digital Democracy, edited by Cynthia Alexander. Toronto: Oxford University Press, 1998.
- 'International Plug 'n Play? Citizen Activism, the Internet, and Global Public Policy.' International Studies Perspectives 1, no. 3 (December 2000): 255–72.

Denning, Dorothy, and Peter J. Denning. *Internet Besieged: Countering Cyber-space Scofflaws.* New York: Addison Wesley, 1998.

Der Derian, James. 'Virtual Security: Technical Oversight, Simulated Foresight, and Political Blindspots in the Infosphere.' In *Cyberspace and Outer Space: Transitional Challenges for Multilateral Verification in the 21st Century,* edited by J. Marshall Beier and Steven Mataija, 105–30. Proceedings of the Fourteenth Annual Ottawa NACD Verification Symposium. Toronto: Centre for International and Security Studies, York University, 1997.

– 'Global Events, National Security, and Virtual Theory.' *Millennium Journal of International Studies* 30, no. 3 (2001): 669–90.

– *Virtuous War: Mapping the Military-Industrial-Media-Entertainment Network.* Boulder, CO: Westview Press, 2001.

Deutsch, Karl. *Nationalism and Social Communication: An Inquiry into the Foundations of Nationality,* 2nd ed. Cambridge, MA: MIT Press, 1969.

– *The Nerves of Government: Models of Political Communication and Control.* New York: Free Press, 1969.

Dewitt, David, David Haglund, and John Kirton, eds. *Building a New Global Order: Emerging Trends in International Security.* Toronto: Oxford University Press, 1997.

Dixit, J.N. *An Afghan Diary: Zahir Shah to Taliban.* Delhi: Konark Publishers Pvt. Ltd., 2000.

Dourish, Paul. *Where the Action Is: The Foundations of Embodied Interaction.* Cambridge, MA: MIT Press, 2001.

Drake, William J., ed. *The New Information Infrastructure: Strategies for U.S. Policy.* Washington, DC: Brookings Institution, 1995.

Drezner, Daniel. 'The Global Governance of the Internet: Bringing the State Back In.' *Political Science Quarterly* 119, no. 3 (Fall 2004): 477–98.

Duncombe, Stephen, ed. *Cultural Resistance Reader.* London: Verso, 2002.

Edwards, David B. 'Print Islam: Media and Religious Revolution in Afghanistan.' *Anthropological Quarterly* 68 (July 1995): 171–84.

– 'Learning from the Swat Pathans: Political Leadership in Afghanistan, 1978–97.' *American Ethnologist* 25, no. 4 (November 1998): 712–28.

El-Nawawy, Mohammed, Adel Iskandar, Adel Iskander, Mohammed El-Nawaway, and Adel Iskandar Farag. *Al Jazeera: How the Free Arab News Network Scooped the World and Changed the Middle East.* Boulder, CO: Westview Press, 2002.

Ellenbogen, Gustavo Gorriti, Robin Kirk, and Gustavo Gorriti. *The Shining Path: A History of the Millenarian War in Peru.* Chapel Hill: University of North Carolina Press, 1999.

Emadi, Hafizullah. 'Exporting Iran's Revolution: The Radicalization of the

Shiite Movement in Afghanistan.' *Middle Eastern Studies* 31 (January 1995): 1–12.

'Enter the Taleban: The Road to Koranistan.' *The Economist*, 5 October 1996, 21–2.

'Fight for the Capital.' *The Economist*, 11 March 1995, 36.

Flindell Klarén, Peter. *Peru: Society and Nationhood in the Andes*. New York: Oxford University Press, 2000.

Foucault, Michel. 'Security, Territory, and Population.' In *Michel Foucault: Ethics, Subjectivity and Truth*, edited by Paul Rabinow, 67–71. New York: New Press, 1994.

Fraser, T.G., ed. *The Irish Parading Tradition: Following the Drum*. London: Macmillan 2000.

Frederick, Howard. *Global Communication and International Relations*. Belmont, CA: Wadsworth, 1993.

Geertz, Clifford. *The Interpretation of Cultures*. New York: Basic Books, 1973.

Girard, Michel, ed. *Individualism and World Politics*. New York: St Martin's Press, 1999.

Goldsmith, Ben R. 'A Victory to Fear or a Source of Hope?' *The World Today*, July 1997, 182–4.

Goldstein, Joshua. *International Relations*, 4th ed. New York: Longman, 2001.

Graham, Brian. 'The Imagining of Place: Representation and Identity in Contemporary Ireland.' In *In Search of Ireland: A Cultural Geography*, edited by Brian Graham, 192–212. London: Routledge, 1997.

– *In Search of Ireland: A Cultural Geography*. London: Routledge, 1997.

Gray, Chris Hables. *Postmodern War: The New Politics of Conflict*. New York: Guilford Press, 1997.

Grover, Verinder, ed. *Afghanistan : Government and Politics*. New Delhi: Deep and Deep Publications, 2000.

Haliday, Fred, and Zahir Tanin. 'The Communist Regime in Afghanistan 1978–1992.' *Europe/Asia Studies* 50, no. 8 (December 1998): 1357–80.

Hamelink, Cees J. *The Politics of World Communication*. Thousand Oaks, CA: Sage, 1994.

Hampson, Fen Osler. *Madness in the Multitude: Human Security and World Disorder*. Toronto: Oxford University Press, 2002.

Hampson, François J. *Incitement and the Media: Responsibility of and for the Media in the Conflicts in the Former Yugoslavia*. Human Rights Centre, University of Essex, Papers in the Theory and Practice of Human Rights no. 3, 1993.

Haraway, Donna, *Modest_Witness@Second_Millennium.FemaleMan©_Meets_Oncomous e*^{TM}*: Feminism and Technoscience*. New York: Routledge, 1997.

Harpviken-Berg, Kristian. 'The Taleban Threat.' *Third World Quarterly* 20, no. 4 (August 1999): 861–70.

Held, David, and Anthony McGrew, eds. *The Global Transformations Reader.* Cambridge: Polity Press, 2000.

Held, David, Anthony McGrew, David Goldblatt, and Jonathan Perraton. *Global Transformations: Politics, Economics and Culture.* Stanford, CA: Stanford University Press, 1999.

Heng, Geraldine. '"A Great Way to Fly": Nationalism, the State, and the Varieties of Third-World Feminism.' In *Feminist Genealogies, Colonial Legacies, Democratic Futures,* edited by M. Jacqui Alexander and Chandra Talpade Mohanty. New York: Routledge, 1997.

Hennessey, Thomas. *Dividing Ireland: World War I and Partition.* London: Routledge, 1998.

Henry, Ryan, and Edward Peartree, eds. *The Information Revolution and International Security.* Washington, DC: CSIS Press, 1998.

Herzstein, Robert E. *The War That Hitler Won: The Most Infamous Propaganda Campaign in History.* New York: Putnam, 1978.

Hill, Kevin A., and John E. Hughes. *Cyberpolitics: Citizen Activism in the Age of the Internet.* Lanham, MD: Rowman and Littlefield, 1998.

Hoffman, Bruce. *Inside Terrorism.* London: Victor Gollancz, 1998.

Hyman, Anthony. 'Afghanistan in Perspective.' *Asian Affairs* 27 (October 1996): 285–95.

Ignatieff, Michael. *Virtual War: Kosovo and Beyond.* New York: Henry Holt, 2000.

'Immorality in Afghanistan.' *The Economist,* 8 March 1997, 36.

James, Edwyn. 'Learning to Bridge the Digital Divide: Computers Alone Are Not Enough to Join the E-economy. Digital Literacy Is Essential Too.' *OECD Observer,* 14 January 2001.

Jarman, Neil. *Material Conflicts: Parades and Visual Displays in Northern Ireland.* Oxford: Berg, 1997.

Jarman, Neil, and Dominic Bryan. 'Green Parades in an Orange State: Nationalist and Republican Commemorations and Demonstrations from Partition to the Troubles, 1920–70.' In *The Irish Parading Tradition: Following the Drum,* edited by T.G. Fraser, 95–110. London: Macmillan, 2000.

Jasani, Bhapendra, and Gotthard Stein. *Commercial Satellite Imagery.* Berlin: Springer-Verlag, 2002.

Jordan, Tim. *Cyberpower: The Culture and Politics of Cyberspace and the Internet.* London: Routledge, 1999.

Jordon, Tim, and Adam Lent, eds., *Storming the Millennium: The New Politics of Change.* London: Lawrence and Wishart, 1999.

Kahin, Brian, and Charles Nesson, eds. *Borders in Cyberspace: Information Policy and the Global Information Infrastructure*. Cambridge, MA: The MIT Press, 1997.

Kalathil, Shantli, and Taylor Boas. *Open Networks, Closed Régimes: The Impact of the Internet on Authoritarian Rule*. Washington, DC: Carnegie Endowment for International Peace, 2003.

Katz, James E., and Ronald E. Rice. *Social Consequences of Internet Use: Access, Involvement, and Interaction*. Cambridge, MA: MIT Press, 2002.

Katz, Jonathan Ned. *The Invention of Heterosexuality*. New York: Plume, 1995.

Kearney, John. *Postnationalist Ireland: Politics, Culture, Philosophy*. London: Routledge, 1997.

Keating, Michael. 'Women's Rights and Wrongs.' *The World Today* 53 (January 1997): 11–12.

Keohane, Robert O., and Joseph S. Nye, Jr. 'Power and Interdependence in the Information Age.' *Foreign Affairs* 77, no. 5 (September-October 1998): 81–92.

Keppel, Gilles. *The War for Muslim Minds: Islam and the West*. Cambridge, MA: Belknap Press, 2004.

Khalilzad, Zalmay. 'Afghanistan in 1994: Civil War and Disintegration.' *Asian Survey* 35 (February 1995): 147–52.

– 'Afghanistan in 1995: Civil War and a Mini-Great Game.' *Asian Survey* 36 (February 1996): 190–5.

– 'Anarchy in Afghanistan.' *Journal of International Affairs* 51 (Summer 1997): 37–56.

Kinsman, Gary. *The Regulation of Desire*. Montreal: Black Rose Books, 1996.

Krasner, Stephen. 'Global Communications and National Power: Life on the Pareto Frontier.' *World Politics* 43, no. 3 (1991): 336–66.

Kroker, Arthur. *Technology and the Canadian Mind: Innis, McLuhan, Grant*. New York: Palgrave Macmillan, 1985.

– *The Possessed Individual: Technology and New French Theory*. New York: St Martin's Press, 1992.

Kroker, Arthur, and David Cook. *The Postmodern Scene: Excremental Culture and Hyper-Aesthetics*. New York: St Martin's Press, 1987.

Kroker, Arthur, and Marilouise Kroker. *Body Invaders: Panic Sex in America*. New York: St Martin's Press, 1987.

– *Hacking the Future: Stories for the Flesh-Eating 90s*. New York: Palgrave Macmillan, 1996.

– *Digital Delirium*. New York: Palgrave Macmillan, 1997.

– *The Last Sex: Feminism and Outlaw Bodies*. New York: St Martin's Press, 2001.

– *The Will to Technology and the Culture of Nihilism*. Toronto: University of Toronto Press, 2003.

Kroker, Arthur, Marilouise Kroker, and David Cook. *Panic Encyclopedia: The Definitive Guide to the Postmodern Scene*. New York: St Martin's Press, 1989.

Kroker, Arthur, and Bruce Sterling. *Spasm: Virtual Reality, Android Music and Electric Flesh*. New York: St Martin's Press, 1993.

Kroker, Arthur, and Michael Weinstein. *Data Trash: The Theory of the Virtual Class*. New York: St Martin's Press, 1994.

Lasswell, Harold. *Propaganda Technique in World War I*. Cambridge, MA: MIT Press 1971.

Lasswell, Harold, and Satish K. Arora. *Political Communication: The Public Language of Political Elites in India and the United States*. New York: Holt, Rinehart and Winston, 1969.

Lasswell, Harold, and Dorothy Blumenstock. *World Revolutionary Propaganda: A Chicago Study*. Westport, CT: Greenwood Press, 1970.

Latham, Robert, ed. *Bombs and Bandwidth: The Emerging Relationship between Information Technology and Security*. New York: New Press, 2003.

Lawson, George. 'Truth or Dare: Truth Commissions between Old and New Nations.' *Open Democracy* (21 November 2002).

Levy, B., and P.T. Spiller, eds. *Regulations, Institutions, and Commitment: Comparative Studies of Telecommunications*. New York: Cambridge University Press, 1996.

'The Lion Flees, and Afghanistan Turns Uglier.' *The Economist*, 1 April 2000, 38.

'Living with the Taleban.' *The Economist*, 24 July 1999, 39–40.

Lloyd, David. *Ireland after History*. Cork, Ireland: Cork University Press, 1999.

Lovnik, Geert. *Dark Fiber: Tracking Critical Internet Culture*. Cambridge, MA: MIT Press, 2002.

Ma'aroof, Mohammad, Khalid. *Afghanistan in World Politics: A Study of Afghan–US Relations*. New Delhi: Gyan Publishing House, 1987.

– *Afghanistan and Super Powers*. New Delhi: Commonwealth Publishers, 1990.

– *United Nations and Afghanistan Crisis*. New Delhi: Commonwealth Publishers, 1990.

MacKinnon, Mark. 'Progress Lagging for Afghan Women.' *Globe and Mail*, 6 August 2003, A10.

MacRaild, Donald M. '"The Bunkum of Ulsteria": The Orange Marching Tradition in Late Victorian Cumbria.' In *The Irish Parading Tradition*, edited by T.G. Fraser, 44–59. London: Macmillan, 2000.

Manguel, Alberto. *Reading Pictures: A History of Love and Hate*. New York: Random House, 2000.

Manovich, Lev. *The Language of New Media*. Cambridge, MA: MIT Press, 2001.

Mariátegui, José Carlos. *Seven Interpretive Essays on Peruvian Reality*. Austin: University of Texas Press, 1988.

Massumi, Brian. *Parables for the Virtual: Movement, Affect, Sensation.* Durham, NC: Duke University Press, 2002.

Matelski, Marilyn. *Vatican Radio: Propagation by the Airwaves.* Westport, CT: Praeger Publishers, 1995.

Matinuddhin, Kamal. *The Taliban Phenomenon: Afghanistan 1994–1997.* New Delhi: Lancer Publishers and Distributors, 2000.

McChesney, Robert W. *Rich Media, Poor Democracy: Communication Politics in Dubious Times.* New York: New Press, 1999.

McCrank, Adele, and Pat Gowens. 'Global Women's Strike 2000.' *Off Our Backs,* March 2000, 6.

McLuhan, Eric, and Frank Zingrone, eds. *Essential McLuhan.* Concord, ON: Anansi Press, 1995.

Meikle, Graham. *Future Active: Media Activism and the Internet.* New York: Routledge, 2002.

Metzl, Jamie Frederic. 'Rwandan Genocide and the International Law of Radio Jamming.' *American Journal of International Law* 91, no. 4 (October 1997): 628. http://www.asil.org/ajil/radio.htm (accessed 23 August 2003).

Miller, David. *On Nationality.* Oxford: Clarendon Press, 1995.

Mirabeau, Comte de. *Les États-généraux.* Le Jay.

Moghadam, Valentine M. *Revolution En-gendered: Women and Politics in Iran and Afghanistan.* Toronto: York University, 1990.

– 'Revolution, Religion, and Gender Politics: Iran and Afghanistan Compared.' *Journal of Women's History* 10, no. 4 (Winter 1999).

Monin, Lydia, and Andrew Gallimore. *The Devil's Garden: A History of Landmines.* London: Pimlico, 2002.

Moody, Bella, ed. *International and Development Communications: A Twenty-First Century Perspective.* Thousand Oaks, CA: Sage, 2003.

Moriarty, Gerry. 'Competition Proposal for Northern Ireland Flag.' *Irish Times,* 20 July 2000.

Mosco, Vincent, and Janet Wasko, eds. *The Political Economy of Information.* Madison: University of Wisconsin Press, 1988.

Mulligan, Therese, and David Wooters, eds. *1000 Photo Icons: George Eastman House.* Cologne: Taschen, 2002.

Murray, Janet. *Hamlet on the Holodeck: The Future of Narrative in Cyberspace.* New York: Free Press, 1997.

Mussington, David. 'The Proliferation Challenges of Cyberspace.' In *Cyberspace and Outer Space: Transitional Challenges for Multilateral Verification in the 21st Century,* edited by J. Marshall Beier and Steven Mataija, 71–92. Proceedings of the Fourteenth Annual Ottawa NACD Verification Symposium. Toronto: Centre for International and Security Studies, York University, 1997.

Nacos, Brigitte Lebens. *Mass-Mediated Terrorism: The Central Role of the Media in Terrorism and Counterterrorism*. Lanham, MD: Rowman and Littlefield, 2002.

Nairn, Tom. *Face of Nationalism: Janus Revisited*. London: Verso, 1997.

Negroponte, Nicholas. *Being Digital*. New York: Vintage Books, 1995.

Nelson, Mariah Burton. *The Stronger Women Get, the More Men Love Football: Sexism and the American Culture of Sports*. New York: Harcourt Brace, 1994.

Nye, Jr, Joseph S., and William Owens. 'America's Information Edge,' *Foreign Affairs* 75, no. 2 (March/April 1996): 20–36.

O'Leary, B., and J. McGarry. *The Politics of Antagonism: Understanding Northern Ireland*. London: Athlone Press, 1993.

O'Toole, Fintan. *The Lie of the Land: Irish Identities*. London: Verso, 1997.

'Our Standpoints.' 'The Burst of the "Islamic Government" Bubble in Afghanistan,' no. 2 (January 1997): 6–7.

Ozouf, Mona. 'Esprit public.' In *Dictionnaire critique de la Révolution française*, edited by François Furet and Mona Ozouf. Paris: Flammarion, 1988.

Palmer, David Scott, ed. *The Shining Path of Peru*. New York: Palgrave Macmillan, 1994.

Parisi, Laura. 'World Wide Web Reviews: Women and International Organizations.' *Feminist Collections* 20, no. 4 (Summer 1999): 11–12.

Parkinson, Alan F. *Ulster Loyalism and the British Media*. Dublin: Four Courts Press, 1998.

Parks, Lisa, and Shanti Kumar, eds. *Planet TV: A Global Television Reader*. New York: New York University Press, 2003.

Patterson, Henry. 'Northern Ireland Economy.' In *Northern Ireland Politics*, edited by A. Aughey and D. Morrow. London: Longman, 1996.

Payind, Alam. 'Evolving Alternative Views on the Future of Afghanistan: An Afghan Perspective.' *Asian Survey* 33 (September 1993): 922–31.

'Peru Country Report on Human Rights Practices for 1998.' Bureau of Democracy, Human Rights and Labor, U.S. Department of State, 26 February 1998.

'Peruvian Rebels Wage Propaganda War on Internet.' Reuters, 3 January 1997.

Pettman, Jan Jindy. *Worlding Women: A Feminist International Politics*. London: Routledge, 1996.

Plummer, Kenneth. *Telling Sexual Stories: Power, Change and Social Worlds*. London: Routledge, 1995.

Pool, Ithiel de Sola. *Technologies without Boundaries: On Telecommunications in a Global Age*. Cambridge, MA: Harvard University Press, 1990.

'Post Cold War Latin America: In the Eagle's Shadow.' *NACLA Report on the Americas* 35, no. 4 (January-February 2002).

Pratap, Anita. *Island of Blood : Frontline Reports from Sri Lanka, Afghanistan and Other South Asian Flashpoints*. New Delhi: Viking Publications, 2001.

Price, Monroe. *Media and Sovereignty: The Global Information Revolution and Its Challenge to State Power*. Cambridge, MA: The MIT Press, 2002.

Qureshi, M.A., et al. *Science, Technology and Economic Development in Afghanistan*. New Delhi: Navrang Booksellers and Publishers, 1987.

Rao, Vinayak. *International Negotiation: The United Nations in Afghanistan and Cambodia*. New Delhi: Manak Publications Pvt. Ltd., 2001.

Rash, W. *Politics on the Nets: Wiring the Political Process*. New York: W.H. Freeman, 1997.

Rashid, Ahmed. 'Sword of Islam.' *Far Eastern Economic Review* 158 (29 December 1994/5 January 1995): 21–2.

– 'Deceptive Peace: Clam Returns to Kabul but Natural Conflict Unresolved.' *Far Eastern Economic Review* 158 (17 August 1995): 28–9.

– 'Austere Beginning: Taleban's Fundamentalist Crackdown Endangers Aid.' *Far Eastern Economic Review* 159 (17 October 1996): 19.

– 'Heart of Darkness.' *Far Eastern Economic Review* 162 (5 August 1999): 8–12.

– 'The Taleban: Exporting Extremism.' *Foreign Affairs* 78, no 2 (November-December 1999): 22–35.

– *Taliban: Militant Islam, Oil and Fundamentalism in Central Asia*. New Haven, CT: Yale University Press, 2000.

Rawls, John. *A Theory of Justice*. Cambridge, MA: Belknap Press, 1971.

Reidenberg, Joel R. 'Governing Networks and Rule-Making in Cyberspace.' In *Borders in Cyberspace: Information Policy and the Global Information Infrastructure*, edited by Brian Kahin and Charles Nesson. Cambridge, MA: MIT Press, 1997.

Reiner, Liz. 'Websites on Women and Human Rights.' *Feminist Collections* 19, no. 3 (Spring 1998): 13–14.

'Report of the Special Rapporteur on the Independence of Judges and Lawyers, Mr Param Cumaraswamy: Report on the Mission to Peru.' Office of the UN High Commissioner for Human Rights, Geneva, Switzerland, 19 February 1998.

'Revenge of the Pathans.' *The Economist*, (25 February 1995): 30–1.

Richmond, Anthony H. *Global Apartheid: Refugees, Racism, and the New World Order*. Toronto: Oxford University Press, 1994.

Robinson, Piers. *The CNN Effect: The Myth of News, Foreign Policy and Intervention*. London: Routledge, 2002.

Rodgers, Jayne, ed. *Spatializing International Politics: Analyzing NGOs' Use of the Internet*. London: Routledge, 2003.

Rodowick, David Norman. *Reading the Figural, or, Philosophy after the New Media*. Durham, NC: Duke University Press, 2001.

Rolston, Bill. *Politics and Painting: Murals and Conflict in Northern Ireland*. Cranbury, NJ: Associated University Presses, 1991.

– *Drawing Support: Murals in the North of Ireland*. Belfast: Beyond the Pale Publications, 1992.

– *Drawing Support 2: Murals of War and Peace*. Belfast: Beyond the Pale Publications, 1998.

Rolston, Bill, and David Miller, eds. *War and Words: The Northern Ireland Media Reader*. Belfast: Beyond the Pale Publications, 1996.

Rosenau, James. *Turbulence in World Politics: A Theory of Change and Continuity*. Princeton, NJ: Princeton University Press, 1990.

– *Distant Proximities: Dynamics beyond Globalization*. Princeton, NJ: Princeton University Press, 2003.

Rosenau, James, and Ernst-Otto Czempiel, eds. *Governance without Government: Order and Change in World Politics*. New York: Cambridge University Press, 1992.

Rosenau, James, and David Johnson. 'Information Technologies and Turbulence in World Politics.' In *Technology, Development, and Democracy: International Conflict and Cooperation in the Information Age*, edited by Julianne Emmons Allison, 55–78. Albany: SUNY Press, 2002.

Rosenau, James, and J.P. Singh. *Information Technologies and Global Politics: The Changing Scope of Power and Governance*. Albany: SUNY Press, 2002.

Rosenberg, Tina. 'Designer Truth Commissions.' *New York Times*, 9 December 2001.

Ruane, Joseph, and Jennifer Todd. *The Dynamics of Conflict in Northern Ireland*. Cambridge: Cambridge University Press, 1996.

Rubin, Barnett R. 'Post-Cold War State Disintegration: The Failure of International Conflict Resolution in Afghanistan.' *Journal of International Affairs* 46 (Winter 1993): 469–92.

Rucht, Dieter. 'Distant Issue Movements in Germany: Empirical Descriptions and Theoretical Reflections.' In *Globalizations and Social Movements: Culture, Power, and the Transnational Public Sphere*, edited by John Guidry, Michael Kennedy, and Mayer Zald, 76–108. Ann Arbor: University of Michigan Press, 2001.

Ryan, Marie-Laure. *Narrative across Media: The Languages of Storytelling*. Lincoln: University of Nebraska Press, 2004.

Said, Edward. *Covering Islam: How the Media and the Experts Determine How We See the Rest of the World*. New York: Panthoen, 1981.

– *After the Last Sky*. New York: Columbia University Press, 1998.

Sassen, Saskia. *Globalization and Its Discontents: Essays on the New Mobility of People and Money*. New York: The New Press, 1998.

Schechter, Danny. *Media Wars: News at a Time of Terror.* Lanham, MD: Rowman and Littlefield, 2003.

Schiller, Dan. 'How to Think about Information.' In *The Political Economy of Information*, edited by Vincent Mosco and Janet Wasco, 27–43. Madison: University of Wisconsin Press, 1988.

– 'Télécommunication, les échecs d'une révolution.' *Le Monde Diplomatique* 592 (July 2003): 28–9.

Shapiro, Andrew. *The Control Revolution: How the Internet Is Putting Individuals in Charge and Changing the World We Know.* New York: Public Affairs, 1999.

Skidmore, Thomas E., and Peter H. Smith. *Modern Latin America*, 5th ed. New York: Oxford University Press, 2001.

Slack, R.S., and R.A. Williams. 'The Dialectics of Place and Space: On Community in the "Information Age."' *New Media and Society* 2, no. 3 (2000): 313–34.

Smith, Anthony. 'Television Coverage of Northern Ireland.' In *War and Words: The Northern Ireland Media Reader*, edited by Bill Rolston and David Miller, 22–37. Belfast: Beyond the Pale Publications, 1996.

Smith, Graeme. 'Pentagon Downed Web Site, Al-Jazeera Editor Says.' *Globe and Mail*, 29 March 2003, A6.

Smith, Michael P., and Luis Eduardo Guarnizo, eds. *Transnationalism from Below.* New Brunswick, NJ: Transaction Publishers, 1998.

Sontag, Susan. *On Photography.* New York: Picador, 2001.

– *Regarding the Pain of Others.* New York: Farrar Straus & Giroux, 2002.

Sreedhar, Sinha, et al. *Taliban and the Afghan Turmoil: The Role of USA, Pakistan, Iran and China.* Mumbai: Himalaya Publishing House, 1997.

Sreedhar, Sinha, and Mahendra Ved. *Afghan Turmoil: Changing Equations.* Mumbai: Himalaya Publishing House, 1998.

Starn, Orin, Robin Kirk, and Carlos I. Degregori, eds. *The Peru Reader: History, Culture, Politics.* Durham, NC: Duke University Press, 1995.

Stern, Steve J., ed. *Shining and Other Paths: War and Society in Peru, 1980–1995.* Durham, NC: Duke University Press, 1998.

Stevenson, Nick. *The Transformation of the Media: Globalisation, Morality and Ethics.* London: Longman, 1999.

Stobdan, P. *Afghan Conflict and India.* New Delhi: Institute for Defence Studies and Analysis, 1998.

'Taleban Defeated.' *The Economist*, 18 March 1995, 38.

'The Taleban Eye the Big Prize.' *The Economist*, 28 October 1995, 38.

Taylor, John. *Body Horror: Photojournalism, Catastrophe and War.* New York: New York University Press, 1998.

Taylor, Peter, and Colin Flint. *Political Geography: World-economy, Nation-state and Locality*, 4th ed. Harlow, Essex: Prentice-Hall, 2000.

Tepperman, Jonathan D. 'Truth and Consequences.' *Foreign Affairs* 81, no. 2 (March/April, 2002): 128–56.

Thomas , Timothy L. 'Al Qaeda and the Internet: The Danger of "Cyberplanning."' *Parameters* (Spring 2003): 112–23.

Thompson, Mark. *Forging War: The Media in Serbia, Croatia, and Bosnia Hercegovina*. Luton, UK: University of Luton Press, 2003.

Thorton, T.P. 'Terror as a Weapon of Political Agitation.' *Internal War: Problems and Approaches*, edited by Harry Eckstein, 78–82. New York: Free Press, 1964.

Thussu, Daya Kishan, ed. *Electronic Empires: Global Media and Local Resistance*. London: Arnold, 1998.

Tomsen, Peter. 'A Chance for Peace in Afghanistan: The Taleban's Days Are Numbered.' *Foreign Affairs* 79, no. 1 (January/February 2000): 179–82.

Toulouse, Chris, ed. *The Politics of Cyberspace*. London: Routledge, 1998.

Truth and Reconciliation Commission. *Final Report*, 28 August 2003.

Turkle, Sherry. *Life on the Screen: Identity in the Age of the Internet*. New York: Touchstone, 1995.

'Unban the Taleban.' *The Economist* (24 July 1999), 19.

UN Security Council Resolution 1325 (2000), 31 October 2000.

Van Kley, Dale K. 'New Wine in Old Wineskins: Continuity and Rupture in the Pamphlet Debate of the French Revolution.' *French Historical Studies* 17, no. 2 (1991): 447–65.

Virilio, Paul. *Open Sky*, translated by Julie Rose. London: Verso, 1997.

– *Ground Zero*. London: Verso, 2002.

Vogel, Tom, Matt Moffett, and Jed Sandberg. 'Tupac Amaru's Web Page Is Hot Spot on the Internet.' *Wall Street Journal*, 6 January 1997.

'The Volley from the Valley.' *The Economist*, 2 August 1997, 33–4.

Wang, Nan. 'Fresh Fighting Erupts in Afghanistan.' *Beijing Review* 42, no. 34 (23 August 1999): 12.

Warikoom, K., ed. *The Afghanistan Cauldron*. New Delhi: Bhavana Books and Prints, 2001.

Warikoom K., Uma Singh, and A.K. Ray. *Afghanistan Factor in Central and South Asian Politics*. New Delhi: Trans Asia Informatics, 1994.

Weber, Max. *The Methodology of Social Sciences*. New York: Free Press, 1949.

Wilkin, Peter. *The Political Economy of Global Communication*. London: Pluto Press, 2001.

Wilkinson, Paul. *Terrorism and the Liberal State*. London: Macmillan, 1977.

Winterson, Jeannette. *Art Objects: Essays on Ecstasy and Effrontery*. Toronto: Vintage, 1995.

'With Gun and Koran.' *The Economist*, 4 February 1995.

Wolcott, Peter. 'Introducing the Global Diffusion of the Internet Series.' *Communications of the Association for Information Systems* 11 (2003): 555–9.

Women, Peace and Security: Study Submitted by the Secretary-General Pursuant to Security Council Resolution 1325 (2000). UN, 2002.

Woods, Oona. *Seeing Is Believing: Murals in Derry.* Derry: Guildhall Press, 1995.

Yu, Sylvia. 'Who Cares for Sex Slaves?' *Globe and Mail*, 17 August 2001, A11.

Yuri the Yaba. 'Yabasta: Mobilizing Global Citizenship through Mass Direct Action.' In *Global Uprising: Confronting the Tyrannies of the 21st Century*, edited by Neva Welton and Linda Wolf. Gabriola Island, BC: New Society Publishers, 2001.

Zalmay, Khalilzad, and John White, eds. *Strategic Appraisal: The Changing Role of Information in Warfare.* Santa Monica, CA: RAND, 1999.

Zelizer, Barbie. *Remembering to Forget: Holocaust Memory through the Camera's Eye.* Chicago: University of Chicago Press, 1998.

Zizek, Slavoj. *On Belief.* London: Routledge, 2001.

Zulaika, Joseba, and William A. Douglass. *Terror and Taboo: The Follies, Fables, and Faces of Terrorism.* New York: Routledge, 1996.

Websites

'About RAWA ...' http://www.rawa.org (accessed 22 August 2001).

'Afghanistan Ends Opium Poppy Cultivation.' *ODCCP Update*, June 2001. http://www.odccp.org/newsletter_2001–06–30_1_page002.html (accessed 20 August 2001).

Agre, Phil. 'Some Thoughts about Political Activity on the Internet,' August 1996. http://www.ucaqld.com.au/news/4political/activity.html (accessed 3 August 1999).

Amnesty International. 'Annual Report 2001 – Covering Events from January–December 2000 Afghanistan.' http://www.amnesty.org.

Anti-Defamation League. http://www.adl.org/.

Article 19: Global Campaign for Free Expression. http://www.article19.0rg/.

Asaravala, Amit. 'College Questioning Site's Link.' *Wired News*, 28 September 2002. http://www.wired.com/news/politics/0,1283,55450,00.html (accessed 18 August 2003).

Baudrillard, Jean. 'Global Debt and Parallel Universe.' 1997. http://www._ctheory.com/e31-global_debt.html.

BBC News. 'Taleban Smash TVs.' 30 July 1998, http://news.bbc.co.uk/hi/english/world/south_asia/newsid_142000/1423 52.stm (accessed 26 March 2005).

- 'Football Tour Cut Short.' 17 July 2000. http://news.bbc.co.uk/1/hi/world/south_asia/837334.htm (accessed 26 March 2005).
- 'Taleban Rules Out Lifting TV Ban,' BBC News, 23 July 2000, http://news.bbc.co.uk/hi/english/world/south_asia/newsid_847000/847408.stm (accessed 26 March 2005).
- 'India's Digital Divide,' 25 May 2003, http://news.bbc.co.uk/2/hi/programmes/from_our_own_correspondent/2 932758.stm (accessed 9 August 2003).
- 'Al-Jazeera Hacker Pleads Guilty.' BBC News, 13 June 2003. http://news.bbc.co.uk/2/hi/americas/2987342.stm (accessed 25 July 2003).
Bimber, Bruce. 'The Internet and Political Transformation.' http://www.sscf.ucsb.edu/~survey1/poltran2.htm (accessed 23 December 1996).
Bonchek, Mark S. 'Grassroots in Cyberspace: Using Computer Networks to Facilitate Political Participation,' MIT Artificial Intelligence Laboratory Working Paper 95–2.2, presented at 53rd Annual Meeting of the Midwest Political Science Association, Chicago, IL. http://www.ai.mit.edu/people/msb/pubs/grassroots.html (accessed 6 April 1995).
Bonchek, Mark, and Jock Gill. 'The Internet and Retail Politics.' http://www.casti.com/gill/presentations/essay0296.html (accessed 10 February 1996).
Chechen Republic. http://www.amina.com/ (accessed 8 August 2003).
CIA. *The World Fact Book 2002*. http://www.cia.gov/cia/publications/factbook/geos/af.html (accessed 9 August 2003).
Clark, Kate. 'The Taleban's Drug Dividend.' *BBC News*, 14 June 2000. http://news.bbc.co.uk/hi/english/world/south_asia/newsid_783000/783268.stm (accessed 26 March 2005).
Clarke, Roger. 'Encouraging Cyberculture.' Address to CAUSE in Australasia '97, Melbourne. http://www.anu.edu.au/people/Roger.Clarke/II/EncoCyberCulture.html (accessed 13–16 April 1997).
Comisión de la Verrdad y Reconciliación. *Final Report*, 28 August 2003. http://www.cverdad.org.pe/ingles/ifinal/conclusiones.php. (accessed 14 December 2004).
Dartnell, Michael Y. 'HTML (HyperText Markup Language) as Needlepoint.' *CTheory* 22, no. 3, article 77 (10 November 1999). http://ctheory.net/text_file.asp?pick=120.
- 'Hyperterrorism: A New Form of Globalized Conflict.' *CTheory* (26 September 2001). http://www.ctheory.net/text_file.asp?pick=305.
- 'The Electronic Starry Plough: The Enationalism of the Irish Republican Socialist Movement (IRSM).' *First Monday* 6, no. 12 (December 2001). http:firstmonday.org/issues/issue6_12/dartnell/index.html.

Denning, Dorothy. 'Activism, Hacktivism and Cyberterrorism: The Internet as a Tool for Influencing Foreign Policy.' www.totse.com/en/technology/cyberspace_the_new_frontier/cyberspr.html (accessed 27 March 2005).

Economic and Social Research Institute. http://www.esri.ie/content.cfm?t=Irish%20Economy&mld=4 (accessed 22 December 2004).

Electronic Iraq. http://electroniciraq.net/.

'E-Stats.' U.S. Department of Commerce, U.S. Census Bureau, Economic and Statistics Adminisration, 18 March 2002. www.census.gov/estats (accessed 5 November 2002).

Guardian, The. http://www.guardian.co.uk.

'Global Petition Campaign for the 10th Anniversary of June 4th Tianamen.' http://www.june4.0rg. (accessed 28 March 2000).

Gluck, Caroline. 'South Korea's Web Guerrillas.' BBC News, 12 March 2003. http://news.bbc.co.uk/2/hi/asia-pacific/2843651.stm.

Golumbia, David. 'Hypercapital,' *Postmodern Culture* 7.1. http://www.english.upenn.edu/~dgolumbi/pmc/hypercapital.html, September 1996.

'Hackers Cripple al-Jazeera Sites.' BBC News, 27 March 2003. http://news.bbc.co.uk/2/hi/technology/2893993.stm (accessed 30 March 2003).

Held, David. 'Violence, Law and Justice in a Global Age.' *After Sept. 11 Archive*, Social Science Research Council, 2001. http://www.ssrc.org/sept11/essays/held.htm (accessed 16 August 2003).

Human and Constitutional Rights Resource Page. www.hrcr.org/docs/frenchdec.html.

Human Rights Watch. *HRW World Report 2000.* http://www.hrw.org/wr2k/Asia.htm#Afghanistan (accessed 14 August 2001).

– 'Afghanistan: Crisis of Impunity, The Role of Pakistan, Russia, and Iran in Fueling the Civil War' 13, no. 3 (C) (July 2001). http://www.hrw.org/reports/2001/afghan2/Afghan0701.htm (accessed 14 August 2001).

– 'Peru Confronts a Violent Past: The Truth Commission Hearings in Ayacucho.' 2002. http://www.hrw.org/americas/peru/.

– 'Essential Background: Overview of Human Rights Issues in Peru,' 31 December 2003. http://hrw.org/english/docs/2004/01/21/peru6988_txt.htm.

– 'Human Rights Watch World Report 1997 – Peru.' http://www.hrw.org/reports/1997/WR97/AMERICAS-08.htm#P387_I64464 (accessed 14 December 2004).

– 'Human Rights Watch World Report 1998 – Peru.' http://www.hrw.org/worldreport/Americas-09.htm#P902_I92900 (accessed 14 December 2004).

International Press Institute. 'World Press Freedom Review – Peru.' http://www.freemedia.at/wpfr/Americas/peru.htm (accessed 14 December 2004).

International Telecommunication Union. http://www.itu.int/ITU-D/ict/statistics/ (accessed 13 August 2003).

Irish Republican Activist Radio. 'IRAradio.com.' http://www.iraradio.com/ (accessed 28 March 2000).

IRSM. 'The Broad Front.' http:www.irsm.org/general/history/broadfront.htm (accessed 25 July 2000).

– 'Capitalism.' http:www.irsm.org/general/history/capitalism.htm (accessed 25 July 2000).

– 'Irish Republican Socialist Movement – 20 years of Struggle.' http:www.irsm.org/general/history/irsm20yr.htm (accessed 25 July 2000).

– 'James Connolly and Irish Freedom.' http:www.irsm.org/general/history/jc&irishfreedom.htm.

– 'Loyalism.' http:www.irsm.org/general/history/loyalism.htm (accessed 25 July 2000).

– 'The Road to Revolution.' http://www.irsm.org/general/history/rtrinireland.htm (accessed 25 July 2000).

– 'What Is Irish Republican Socialism?' http://www.irsm.org/general/history/whatis.htm (accessed 8 August 2003).

– 'What Is National Liberation?' http://www.irsm.org/general/history/whatisnatlib.htm (accessed 25 July 2000).

– 'Why the IRSP?' http://www.irsm.org/general/history/whyirps.htm (accessed 25 July 2000).

– 'Women in Ireland.' http:www.irsm.org/general/history/women.htm (accessed 25 July 2000).

'King Billy Mural.' http://cain.ulst.ac.uk/bibdbs/murals/slide1.htm#1 (accessed 29 March 2000).

McAllester, Matthew. 'Peruvian Rebels Go to the Web.' *New York Newsday.* http://www.newsday.com (accessed 8 January 1997).

McLaughlin, W. Sean. 'The Use of the Internet for Political Action by Non-state Dissident Actors in the Middle East.' *First Monday* 8, no. 1 (2003). http://firstmonday.org/issues/issue8_11/mclaughlin/.

'Measuring Electronic Business: Definitions, Underlying Concepts and Measurement Plans.' Thomas L. Mesenbourg, Assistant Director for Economic Programs, Bureau of the Census, 13 October 1999. http://www.ecommerce.gov/ecomnews/e-def.html (accessed 16 November 1999).

Morris, William. 'Art and Socialism.' http://www.marx.org/morris/1884/as/as.html (accessed 2 May 1998).

MRTA. 'How Long Will the Residence Stay Occupied?' (24 December 1996).

- 'Interview with Victor Polay – 1990.'
- 'Interview with an MRTA Leader, Comandante Andres – January 1991.'
- 'Is the MRTA's Action Weakening Fujimori?' (30 December 1996).
- 'Is a Solution in Lima at Hand?' (24 March 1997).
- 'Letter from MRTA Political Prisoners – November 1996.'
- 'The Lives of Political Prisoners in Peru Are in Danger!'
- 'Neo-Liberalism and Globalization.'
- 'The Situation of MRTA Political Prisoners in Peru – May 1996.'
- 'Struggle against Neo-Liberalism!' MRTA communiqué.
- 'What Are the Goals of the Embassy Occupation?' (19 December 1996).

New York Stock Exchange. http://www.nyse.com/ (accessed 28 March 2000).

North County Times. 19 December 2000. http://www.nctimes.com/news/
121900/y.html.

'Northern Ireland Economic Overview.' Northern Ireland Office, October 1997.
http://www.nio.gov.uk/970919.htm (accessed 25 July 2000).

O'Connell, P.J., R.O. O'Donnell, and V. Gash. *Astonishing Success: Economic
Growth and the Labour Market in Ireland.* Dublin: The Economic and Social
Research Institute. http://www.esri.ie/1999_BK_MN_SUM.HTM#
AstonishingSuccess (accessed 25 July 2000).

Packer, George. 'Smart-Mobbing the War.' *New York Times*, 9 March 2003.
www.nytimes.com (accessed 20 April 2003).

Poster, Mark. 'CyberDemocracy: Internet and the Public Sphere.' http://
www.hnet.uci.edu/mposter/writings/democ.html (accessed 1995).

'President Signs Afghan Women and Children Relief Act, Dec. 21, 2001.' White
House press release. http://www.whitehouse.gov/news/releases/2001/
12/20011212–9.html.

Puget Sound Business Journal Online, 24 May 1999. http://seattle.bizjournals
.com/seattle/.

'Puppet soldiers.' http://cain.ulst.ac.uk/bogsideartists/bsprotest2.htm
(accessed 29 March 2000).

'Quarterly Economic Commentary.' Dublin: The Economic and Social Research
Institute, June 2000. http://www.esri.ie/QEC0600.HTM (accessed 25 July
2000).

'RAWA Holds Press Conference after Being Forced to Cancel Its Protest Rally
April 27, 2001.' http://songs.rawa.org/rawa/apr28–01r.htm.

'Regional Action Plan adopted by "Six plus Two" Group in New York on
13 September 2000.' UN ODCCP. http://www.odccp.org/uzbekistan/
actionplan.html (accessed 20 August 2001).

Reporters Without Borders, '1997 Report.' http://www.calvacom.fr/rsf/
RSF_VA/Rapp_VA/Carte_VA/Rapp_VA.html.

- 'Annual Report 2001.' http://www.rsf.fr/uk/home.html (accessed 16 August 2001).
- 'The Enemies of the Internet – Peru.' http://www.rsf.fr/rsf/uk/html/internet/pays_internet/perou.html (accessed 14 December 2004).
- 'Peru – Annual Report 2002.' http://www.rsf.fr.print.php3?id_article=1417 (accessed 14 December 2004).
- 'Peru – Annual Report 2004.' http://www.rsf.fr/article.php3?id_article =10268&Valider=OK (accessed 14 December 2004).

'Republic of Chechnya.' http://www.anima.com (accessed 28 March 2000).

'Revolutionary Association of the Women of Afghanistan' (RAWA). http://www.rawa.org (accessed 28 March 2000).

Scheffler, Mark. 'Scare Tactics: Why Are Liberian Soldiers Wearing Fright Wigs.' *Slate*, 1 August 2003. http://slate.msn.com/id/2086490 (accessed 13 August 2003).

Simon Weisenthal Center http://www.wiesenthal.com/.

Southern Poverty Law Center. http://www.splcenter.org/.

Swett, Charles. 'Strategic Assessment: The Internet.' Office of the Assistant Secretary of Defense for Special Operations and Low-Intensity Conflict, July 1995. www.isoc.org/inet96/proceedings/e1/e1_2.htm.

Tolerance.org. http://www.tolerance.org/.

Tyler, Patrick E. 'A New Power in the Streets.' *New York Times*, 17 February 2003. www.nytimes.com (accessed 20 April 2003).

'Understanding the Digital Economy.' U.S. Department of Commerce. http://www.digitaleconomy.gov/ (accessed 20 January 2001).

'UNHCR's Contribution to a Special Inter-agency Briefing on Pakistan/Afghanistan – Urgent Need for Camp Sites for Afghan Refugees in Pakistan,' 8 February 2001. http://www.unhcr.ch/news/cupdates/0101afg.htm (accessed 15 August 2001).

UNICEF. 'A Humanitarian Appeal for Children and Women, January–December 2001 – Afghanistan.' http://www.unicef.org (accessed 2 January 2002).

USA Patriot Act HR 3162 (Uniting and Strengthening America by Providing Appropriate Tools Required to Intercept and Obstruct Terrorism (USA PATRIOT ACT) Act of 2001). http://www.hqda.army.mil/rio/links/USA%20PATRIOT%20Act%20HR %203162.htm (accessed 25 July 2003).

'US Government Electronic Commerce Policy.' http://www.ecommerce.gov/ (accessed 28 March 2000).

'W.A.R.' (White Aryna Resistance). http://www.resist.com (accessed 28 March 2000).

Ward, Mark. 'Websites Spread al-Qaeda Message.' BBC News, 12 December

2002. http://news.bbc.co.uk/2/hi/technology/2566527.stm (accessed 13 December 2002).

Wellman, Barry, Laura Garton, and Caroline Haythornthwaite. 'Studying Online Social Networks.' *Journal of Computer-Mediated Communication* 3, no. 1 (June 1997). http://jcmc.indiana.edu/vol3/issue1/garton.html (accessed 24 March 2005).

'What the World Thinks in 2002: How Global Publics View: Their Lives, Their Countries, The World, America.' *The Pew Global Attitudes Survey*, The Pew Research Center, 4 December 2002. http://people-press.org/reports/display.php3?ReportID=165 (accessed 30 December 2002).

Wilde, Oscar. 'House Decoration.' Reprinted from *Essays and Lectures by Oscar Wilde*. London: Methuen, 1908. http://www.burrows.com/founders/house.html (accessed 2 May 1998).

'Women's Rights in Afghanistan and Beyond.' April Palmerlee, Senior Coordinator for International Women's Issues, Remarks to MSNBC Subscribers, 5 September 2002. http://www.state.gov/g/wi/rls/13593.htm.

The World Fact Book 2000. www.umsl.edu/services/govdocs/wofact2000/ (accessed 27 March 2005).

'YA BASTA! (Zapatista National Liberation Army).' http://www.ezln.org/ (accessed 29 March 2000).

Index

Digital Futures is a series of critical examinations of technological development and the transformation of contemporary society by technology. The concerns of the series are framed by the broader traditions of literature, humanities, politics, and the arts. Focusing on the ethical, political, and cultural implications of emergent technologies, the series looks at the future of technology through the 'digital eye' of the writer, new media artist, political theorist, social thinker, cultural historian, and humanities scholar. The series invites contributions to understanding the political and cultural context of contemporary technology and encourages ongoing creative conversations on the destiny of the wired world in all of its utopian promise and real perils.

Series Editors:
Arthur Kroker and Marilouise Kroker

Editorial Advisory Board:
Taiaiake Alfred, University of Victoria
Michael Y. Dartnell, York University
Ronald Deibert, University of Toronto
Christopher Dewdney, York University
Sara Diamond, Banff Centre for the Arts
Sue Golding (Johnny de philo), University of Greenwich
Pierre Levy, University of Ottawa
Warren Magnusson, University of Victoria
Lev Manovich, University of California, San Diego
Marcos Novak, University of California, Los Angeles
John O'Neill, York University
Stephen Pfohl, Boston College
Avital Ronell, New York University
Brian Singer, York University
Sandy Stone, University of Texas, Austin
Andrew Wernick, Trent University

Books in the Series:
Arthur Kroker, *The Will to Technology and the Culture of Nihilism: Heidegger, Nietzsche, and Marx*
Neil Gerlach, *The Genetic Imaginary: DNA in the Canadian Criminal Justice System*
Michael Strangelove, *The Empire of Mind: Digital Piracy and the Anti-Capitalist Movement*
Tim Blackmore, *War X: Human Extensions in Battlespace*
Michael Y. Dartnell, *Insurgency Online: Web Activism and Global Conflict*